The

CONTROL
AND
TREATMENT
of
INDUSTRIAL
AND
MUNICIPAL
STORMWATER

Edited by
Peter E. Moffa

VAN NOSTRAND REINHOLD
A Division of International Thomson Publishing Inc.

New York • Albany • Bonn • Boston • Detroit • London • Madrid • Melbourne
Mexico City • Paris • San Francisco • Singapore • Tokyo • Toronto

Contents

Preface

The significance of stormwater pollution was recognized in the early 1970s as a component of combined sewer overflows. Since then, separate urban and industrial stormwater has been identified as a pollution source, requiring its own distinct regulation. In November 1990, EPA issued the stormwater regulations to deal with this problem. However, the size and complexity of this document provides a challenge to even the most dedicated manager. Whereas, on the one hand, the regulations are very comprehensive, individual states have embraced them in varying degrees, from full and complete implementation to marginal participation at best.

This has resulted in misplaced acceptance, with little likelihood of genuine acceptance. This is reinforced by the 1995 Congressional trend to minimize the impact of such requirements, and specifically to grant the states greater leeway in their implementation of the stormwater regulations.

This book presents a historical perspective of the stormwater problem and the legal foundation for these regulations, with emphasis on the philosophy and direction of the federal requirements. Methodologies are provided to assist a municipality, industry, or industrial group in defining the extent of its problem and the actual impact on receiving waters.

Abatement technologies are outlined that range from the simple housekeeping steps or best management practices (BMPs) to the capital-intensive facilities. Sufficient detail is provided to formulate the basis for an effective stormwater pollution prevention plan.

Case histories are presented to provide various perspectives on the problems faced with stormwater pollution. Whether a specific industrial site or a watershed basin approach, each investigation adds to an overview of what different communities have to deal with. This book focuses on available solutions to the problem and actual histories where these solutions have been put to use.

Acknowledgments

This text represents the efforts of several individuals: Thanks to my coauthors for their dedication and perseverance; to the clients I have served over the last several years that have given me the opportunity to work on their stormwater problems; to Leonard T. Wright and Howard M. Goebel for their suggestions and assistance; and, lastly, to my wife and three daughters, Michele, Larisa, and Stephanie, for their support.

About the Authors

Chapter 1

Peter E. Moffa is a principal with Moffa & Associates, Consulting Engineers, with more than 25 years of international experience in stormwater abatement technology. Mr. Moffa has worked on stormwater related projects for over twenty municipalities and several industries, ranging from stormwater modeling to abatement. Mr. Moffa is Editor and prime author of the textbook *Control and Treatment of Combined Sewer Overflows.*

Neal J. Cabral, an Associate of McGuire, Woods, Battle and Boothe, L.L.P., concentrates in environmental issues under the Clean Water Act. He counsels clients on municipal and industrial wastewater discharge permits, including stormwater permitting and compliance issues.

Libby Ford is a Senior Envirnmental Health Engineer with Nixon, Hargrave, Devans and Doyle. Ms. Ford has extensive working knowledge in a broad range of federal, state, and local environmental regulations for the industrial and municipal sector.

Chapter 2

Stephan J. Nix is an associate professor of civil and environmental engineering at The University of Alabama. He has nearly 20 years of experience in urban stormwater modeling and management and has published numerous papers and reports in this area. He was one of the principal contributors to EPA's Storm Water Management Model and has directed a number of sewer system modeling studies.

Chapter 3

Paul L. Freedman is President of Limno-Tech, Inc., a firm specializing in environmental engineering and assessment. He has over 20 years experience in evaluating and modeling stormwater and water quality issues involving over 300 locations in 25 states and provinces. Besides his predominant work for regulated clients, he also provides extensive training, technical support, and model development for EPA and State agencies on stormwater and water quality issues.

David W. Dilks is an Associate Vice President at Limno-Tech. He has 15 years experience in modeling water quality impacts of stormwater and other sources at over 150 sites nationwide.

Chapter 4

Richard Field has been in charge of the U.S. EPA's National Urban Wet-Weather Flow (WWF) Management and Pollution Control R&D Program and in that capacity has gained 25 years of experience in stormwater management and pollution control. He is the U.S. EPA's WWF expert, has been responsible for more than 300 R&D projects (many have been full-scale abatement facilities), and has published more than 400 papers/reports/ book chapters/books.

1

Overview of Stormwater Requirements

BACKGROUND

In 1972, the amendments to the Federal Water Pollution Control Act were most notable in terms of both their national comprehensiveness and their specificity. Virtually all sources of pollution were recognized, with the possible exception of agricultural runoff, ranging from municipal and industrial dry-weather discharges to discharges resulting from wet-weather events. The amendments specified that any point source discharge of a pollutant to surface water without a National Pollutant Discharge Elimination System (NPDES) permit was unlawful. Since 1972 we have gradually moved from the abatement of the most obvious and concentrated forms of wastewater, the dry-weather discharges, to the control of

1

the more nebulous and diverse forms of pollution, the wet-weather discharges.

The first priority in implementing the NPDES program was to require a minimum of secondary treatment for municipal discharges and subject industry to specific pollutant limitations. Having largely accomplished that objective, some municipalities moved on to the abatement of combined sewer overflows (CSOs), which has awakened communities to the problems of stormwater-induced pollution. Combined sewers are defined as sewers designed to convey a mixture of sanitary wastewater and stormwater during wet-weather events. CSOs are addressed through another text, entitled "Control and Treatment of Combined Sewer Overflows." The subject of the following text is stormwater conveyed by separate storm sources, ditches, or any defined conveyance structure that is not designed to carry sanitary wastewater. Industrial as well as municipal stormwater is addressed in this text.

There is another form of pollution, one that will not be addressed in this text: namely, non-point pollution. For the purpose of this text, non-point pollution is pollution that emanates from sources that are not *regulated* as point sources under the NPDES program. It is this type of pollution that has stimulated the need to take a more holistic approach in evaluating pollution abatement requirements, as evidenced by the latest buzzword, "watershed management." As stated in the EPA CSO Control Policy, "Permitting authorities are encouraged to evaluate water pollution control needs on a watershed management basis and coordinate CSO control efforts with other point and non-point source control activities." Urban and industrial stormwater discharges addressed in this book represent a component of such an overall watershed management approach, as illustrated in Figure 1.1.

In 1973, EPA promulgated its first stormwater regulations, in the form of an exemption. Any conveyances

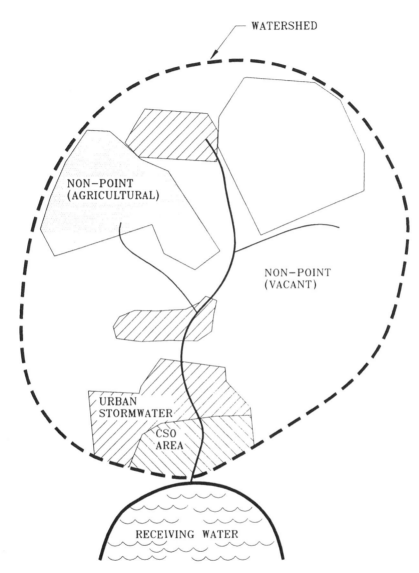

Figure 1.1.   Watershed stormwater components.

carrying stormwater runoff uncontaminated by industrial or commercial activity were exempt from these regulations. EPA reasoned that because of the intermittent, variable, and unpredictable nature of such stormwater, such discharges would be better managed at the local level through non-point source programs (e.g., programs other than NPDES). It was also recognized at that time that the agency would find it difficult issuing individual NPDES permits for the thousands of such non-point source discharges, which would create an overwhelming administrative burden and would distract from the control of municipal and industrial sewage problems that were higher priority at that time. However, the Natural Resources Defense Council (NRDC) brought suit in the U.S. District Court for the District of Columbia, challenging the agency's authority to create such exemptions. The District Court ultimately held that EPA could not exempt discharges identified as point sources from regulations under the NPDES permit program.

Subsequently, EPA issued a rule in 1976 establishing a permit program for all stormwater discharges except for rural runoff uncontaminated by industrial or commercial activity. In 1980, EPA published comprehensive revisions to the NPDES regulations that required the same application information for stormwater point sources as that required for all industrial and commercial process wastewater discharges, including toxic pollutants as identified in the 1977 amendments to the Clean Water Act (CWA). What followed was a series of suits brought by major trade associations and their member companies, a process that culminated in the NPDES Settlement Agreement in 1982. This resulted in stormwater regulations in 1984; however, they too generated considerable controversy and once again suits were filed, resulting in several proposed application requirements for stormwater discharges in 1985.

During EPA's evaluation of appropriate changes, Congress passed the Water Quality Act of 1987 (WQA) which amended the CWA and included specific provisions related to stormwater, most notably in the area of water quality, whereas previous legislation had specifically tar-geted water quantity. The WQA provides that EPA or states with NPDES programs could not require permits for certain stormwater discharges until October 1, 1992,* except for those stormwater discharges that are specifi-cally exempted. The five types of stormwater discharges that were required to obtain a permit prior to October 1, 1992 were as follows:

1.   a discharge for which a permit had been issued prior to February 4, 1987

2.   a discharge associated with industrial activity

3.   a discharge from municipal separate storm sewer systems serving a population of 250,000 or more

4.   a discharge from a municipal separate storm sewer system serving a population of 100,000 or more but less than 250,000

5.   a discharge for which EPA or a state with an approved NPDES program determines that the stormwater discharge contributes to a water quality standard or is a significant contributor of pollutants to the waters of the United States.

The WQA went on to require EPA to promulgate final regulations governing stormwater discharges associated with industrial activity and the above-identified munici-pal-separate storm sewer systems. The conference report accompanying the WQA provided that, after October 1, 1992,* the permit requirements of the CWA were to be restored for municipal-separate storm sewer systems serving a population of fewer than 100,000. This is an

---

*This deadline was extended to October 1, 1994 by subsequent legislation.

important, but often-missed, requirement for the smaller municipalities.

The WQA specified that NPDES permits for stormwater discharges associated with industrial activity must meet essentially all of the applicable provisions including technology and water-quality-based standards. However, the WQA made significant changes to the permit standards for discharges for municipal-separate storm sewers. NPDES permits for such discharges

- may be issued on a system- or jurisdiction-wide basis
- shall include a requirement to effectively prohibit non-stormwater discharges into the storm sewers
- shall require controls to reduce the discharge of pollutants to the maximum extent practicable, including management practices, control techniques, system design, and engineering methods, and such other provisions as EPA or the state determines appropriate for the control of such pollutants.

The EPA, in consultation with the states, was required to conduct two studies on stormwater discharges that would not be required to be permitted prior to October 1, 1992. The first study was to identify those stormwater discharges not required to be permitted, and to determine, to the maximum extent practicable, the nature and extent of pollutants in such discharges. The second study was for the purpose of establishing procedures and methods to control stormwater discharges to the extent necessary to mitigate impacts on water quality. EPA was to have issued by no later than October 1, 1992,* related information that

---

*This deadline was extended to October 1, 1994 by subsequent legislation.

would establish priorities, requirements, and expeditious deadlines.

The WQA also amended the CWA to provide that the EPA shall not require a permit for discharges of stormwater runoff from oil and gas operations that is not contaminated by contact or does not come in contact with any potential pollutant materials such as overburden raw material, intermediate product, finished product, by-product, or waste product located on the site of such operations. The WQA also amended the CWA to exclude agricultural stormwater discharges from stormwater regulations.

EPA has developed a strategy for issuing permits for industrial stormwater discharges. This strategy may serve to establish the rationale for municipal stormwater as well. In many states the strategy will serve only as an underlying basis for the more practical approach of coordinating stormwater permits with the routine renewal of other discharge permits. Nevertheless, the EPA strategy will help in understanding the general perspective of regulatory agencies. A four-tier set of priorities has been established, as follows.

- Tier I: *Baseline Permitting.* One or more general permits will be developed to cover initially the majority of stormwater discharges associated with industrial activity.
- Tier II: *Watershed Permitting.* Facilities within watersheds shown to be adversely impacted by stormwater discharges associated with industrial activity will be targeted for individual or watershed-specific general permits.
- Tier III: *Industry-Specific Permitting.* Specific

---

*This deadline was extended to October 1, 1993 by subsequent legislation.

industry categories will be targeted for individual
or industry-specific general permits.
• Tier IV: *Facility-Specific Permitting.* A variety of
factors will be used to target specific facilities for
individual permits.

The directives of the 1987 WQA were threefold. First,
Congress specifically directed EPA to address stormwater
discharges under the NPDES program, and established
statutory deadlines for the initial phases of the program.
Second, Congress affirmed that stormwater discharges
from industrial sites must be issued NPDES permits, and
that the full panoply of traditional NPDES permit requirements, including technology-based and water-quality-
based standards, must be applied. Third, Congress established slightly different permitting requirements and
standards for municipal stormwater discharges than for
industrial stormwater discharges, including the new
"maximum extent practicable" standard.

Development of this legislation was an arduous task.
An outline of the applicable legislation developed by the
EPA for specific stormwater issues is provided in Appendix A, to provide a framework of the detailed features
contained within the series of four primary Federal mandates.

EPA's NPDES Stormwater Requirements

The 1987 Amendments of the Clean Water Act required
EPA to implement the NPDES permit system to regulate
stormwater discharges. NPDES permits are required from
"point source" stormwater discharges to "waters of the
United States." A "point source" discharge is loosely defined as any discernible conveyance from which pollutants may be discharged. By EPA's definition, "waters of

the United States" include most surface water, based on the potential for commerce, travel, and recreation. These regulations have required both industries and municipalities to develop pollution prevention measures focused on identifying pollution sources and implementing measures that prevent or control the pollution from stormwater discharges.

Industrial

The EPA issued their final rule for NPDES permit applications on November 16, 1990. The rule requires facilities with "stormwater discharge associated with industrial activity" to apply for a stormwater discharge permit. This ruling greatly increased the number and types of facilities under NPDES coverage. The EPA has broadly defined eleven categories of industrial activity in which a facility must be engaged to be subject to the stormwater regulations. The eleven industrial categories are summarized below:

1. heavy manufacturing facilities

2. medium manufacturing facilities

3. mines (active or inactive) and oil and gas facilities

4. hazardous waste treatment storage and disposal facilities

5. landfills, disposal facilities

6. recycling facilities (scrap/salvage yards)

7. steam electric generating facilities

8. selected transportation facilities

9. domestic sewage treatment works

10. construction activities disturbing five or more acres of land

11. other industrial facilities exposed to stormwater

These broad industrial categories were developed to index types of activities that have a similar propensity for stormwater discharge as part of their normal operation. Industries that discharge stormwater into combined sewers are exempt from coverage under this program.

There are three types of NPDES stormwater permit applications available for stormwater discharges: individual, general (NOI), and group permit applications. An individual permit is issued from individual permit applications, whereas a general permit is issued from general and group permit applications. These permits are intended to authorize stormwater discharges associated with industrial activity and prohibit nonstormwater discharge.

Individual permits are issued to a facility on a case-by-case basis, with terms and conditions tailored to the discharge characteristics of the particular facility. The deadline for applying for individual permits was October 1, 1992. This alternative required the discharger to provide detailed information requested on the application and to conduct stormwater sampling. Individual permits offered the applicant a greater degree of control because each application was individually reviewed by the permitting authority and the terms could be negotiated.

An appreciable amount of information was required for an industrial permit application, including a site map, drainage area delineation, description of priority material usage, monitoring data, description of existing controls, and disclosure of spills or leaks of significant materials. Individual permits normally specify effluent limitations and monitoring requirements that the facility must meet.

General permits are similar to individual permits, with the exception that general permits can authorize discharge from more than one facility. General permits offer a less arduous alternative to site-specific permitting for both the discharger and the permitting authority. To be covered under a general permit, a discharger agrees to the terms of the general permit by submitting a brief Notice of Intent (NOI) in lieu of the lengthy individual permit application. The deadline for submission of a NOI for a general permit was October 1, 1992. Industrial facilities starting industrial activity after the October 1, 1992 deadline are required to submit the NOI at least two days prior to facility opening.

General permits prohibit discharges containing hazardous substances established at 40 CFR 117.3 or 40 CFR 302.4. If the permitting agency believes there is a significant risk of stormwater discharge containing a hazardous substance, an individual permit will be required to address this concern adequately.

The focus of a general permit is the development and implementation of stormwater pollution prevention plans (SPPP) using best management practices (BMPs), rather than pollutant-specific limits to control stormwater pollution. These plans do not require review or approval by the permitting authority unless specifically requested.

To reduce and control pollutants contained in stormwater discharges, EPA determined that SPPPs must be developed and implemented for each facility covered by a NPDES permit on a site-specific basis. Detailed SPPPs are required regardless of what type of permit a facility designates. The plan must identify potential pollution sources affecting the quality of stormwater discharges associated with industrial activity. Further, the plan must evaluate and implement BMPs to reduce pollution from stormwater associated with the facility. All SPPP must contain the following information.

1. *Planning and Organization.* Initiation of a Pollution Prevention Team to identify the personnel responsible for development, submission, implementation, and maintenance of the facility's SPPP.

2. *Pollutant Source Assessment.* This is a description of the potential pollution sources from the industrial facility.

3. *BMP Identification.* The SPPP must specify measures and controls to eliminate or reduce stormwater pollution from the industrial facility.

4. *Site Evaluation.* The SPPP must specify measures to ensure compliance with the terms and conditions of the permit and to assess the effectiveness of the plan.

Another option for industrial discharges was to apply for a general permit, using a group permit application.

Group permit applications could be used to generate a customized permit for facilities that have similar process characteristics. Group permit applications were developed to reduce the application requirements of industrial facilities. A group must have at least four members to apply under this designation. The group appoints an administrator who assembles the baseline information, materials stored outside, and the management practices of each member. In addition, a sampling program must be initiated to obtain information regarding the prospective pollutants discharged. For groups with more than 10 members, 10% of the members are required to conduct a sampling program. For groups with four to 10 members, 50% of the members are required to be involved with the sampling program. Upon review of the application by the permitting authority, the group could negotiate the terms of the permit with the permitting authority.

Facilities subject to the annual reporting of EPCRA Section 313 for water priority chemicals are required to develop long-term stormwater control plans and to meet performance-based standards. This special provision requires development of BMPs for storage, processing, and handling associated with the water priority chemical(s). The plan must address preventive maintenance and facility security and provide for employee training of individuals involved in plan implementation. Additionally, the special requirements provided in the control plan are to be reviewed and certified by a Registered Professional Engineer.

Municipal

Under the WQA of 1987, EPA was required to issue NPDES permits to large (serving a population of 250,000 or more) and medium (serving a population of 100,000 or more but less than 250,000) municipal separate sewer systems (MS4s). Under the statute, the permits

- may be issued on a system- or jurisdiction-wide basis
- shall prohibit non-stormwater discharges into the storm sewers
- must require controls to reduce the discharge of pollutants

Regulations define "municipal separate stormwater" as a conveyance or system of conveyances (including roads with drainage systems, municipal streets, catch basins, curbs, gutters, ditches, manmade channels, or storm drains) that is owned or operated by a state or local government entity for collecting or conveying stormwater and is not part of a publicly owned treatment work (POTW). Combined sewer systems or POTWs are excluded from stormwater permit requirements because they should already be permitted under existing NPDES programs. The MS4 permit application is subdivided into 2 parts.

Part 1 of the MS4 permit application provides information on the following elements:

- general information
- legal authority
- source identification
- discharge characterization
- management programs
- fiscal resources

Part 2 of the permit application provides information to supplement Part 1 and proposes a program of structural and nonstructural pollution control measures to reduce discharge of pollutants to the "maximum extent practicable."

The primary focus of the Part 2 application is the proposed Stormwater Management Plan (SWMP). This plan to be implemented during the term of the permit is intended to reduce the discharge of pollutants "to the maximum extent practicable" using management practices, control techniques, and design modifications. These measures shall include discharge characterization by implementation of a physical monitoring program, evaluation of existing control features, development of appropriate abatement measures, and estimation of pollutant loading reductions realized through the implementation of the SWMP.

Construction Sites

On September 9, 1992, EPA issued a general permit covering stormwater discharges associated with construction activities. For a construction site to be covered under the general permit, a discharger must submit an NOI that provides information similar to the industrial NOI. A stormwater pollution prevention plan (SPPP) is required for construction sites prior submission of the NOI. The following elements must be provided in all construction site SPPPs:

1. site description

2. measures and controls

3. maintenance schedule

4. inspection schedule.

Construction site SPPPs are primarily intended to control runoff, stabilize soil, and control sediment that results from construction activities. To ensure that the SPPP is followed, all contractors and any subcontractors must certify in writing that they understand the terms and conditions of the stormwater general permit.

The general permit for construction sites does not require any monitoring or submission of reports unless specifically requested by the permitting authority under special circumstances. The general permit is in effect until a notice of termination has been filed after the site has reached final stabilization.

THE STORMWATER REGULATIONS—THE LEGAL PERSPECTIVE

The evolution of the EPA stormwater mandates has been directed by modifications to the Federal Doctrine and numerous legal challenges to such regulations. This section provides a description of the legal perspective of the development of federal stormwater mandates.

The particular stormwater statute was finalized as the

Water Quality Act (WQA) of 1987. The principal stormwater documents that were developed to meet the WQA directives are cited in this section. Specific stormwater legislation was introduced as regulatory preambles, which include Sections 55, 57, and 58 of the Federal Register (*FR*). The official stormwater regulations derived from the regulatory preambles are contained in section 40 of the Code of Federal Regulations (40 CFR). These documents contain the federal regulations associated with stormwater permitting and control and are frequently cited throughout this section.

In response to the directives of the WQA, EPA extended the NPDES program to address stormwater discharges. Although this program appears confusing at first glance, it is not really very different from EPA's traditional NPDES process wastewater permitting program. Most of the confusion stems from the historical precedent of EPA's establishment of the minimum requirements, and then NPDES states' adoption of programs at least as stringent as EPA's program. EPA's stormwater program employs EPA's general permitting authority, which has been rarely used in the NPDES context, and it authorizes a unique one-time group permit application process. However, the bulk of the stormwater requirements is in the framework of EPA's existing NPDES program. The stormwater requirements establish which types of discharges must get a permit, define permit application criteria, and set out minimum permit terms and conditions, such as required pollutant controls, monitoring requirements, and reporting and recordkeeping requirements.

Industrial

EPA has initially implemented the NPDES permitting program for "stormwater discharges associated with industrial activity" through two major rulemakings: the permit application rulemaking and the general permit rulemaking. These two rulemakings affirm that industrial stormwater permits are simply another type of NPDES permit. Familiar NPDES permit program concepts, such as the definition of *point source,* signatory requirements, and monitoring and recordkeeping requirements, are retained in the industrial stormwater permit regulations. Consequently, if a particular issue arises in the stormwater context, reference may be had to the traditional NPDES regulations and the body of law and guidance interpreting them.

Industrial Facilities Required to Obtain a Stormwater Permit

Under EPA's industrial NPDES stormwater permit program, "stormwater discharges associated with industrial activity" require the filing of permit applications and the attainment of NPDES permits for stormwater discharges. To determine whether a particular industrial facility needs a stormwater permit, two general questions must be addressed: (1) Is the stormwater discharge covered under the CWA at all? and (2) If the discharge is covered under the CWA, is the discharge a "stormwater discharge associated with industrial activity"? We shall address each of these broad questions in turn.

Discharges Covered under the CWA

The CWA applies only to a "discharge" of pollutants from a "point source" into "waters of the United States." (See 33 U.S.C. §§ 1311(a), 1362(12)). Stormwater discharges are no different. Two elements of this definition are particularly important: "point source" and "waters of the United States."

First, the definition of "point source" should be carefully examined. Although you are unlikely to succeed in court on an argument that a process wastewater discharge is not from a point source, there is a possibility that such an argument could be won under certain limited circumstances with discharges of stormwater. The Clean Water Act defines "point source" very broadly as

> any discernible, confined and discrete conveyance, including but not limited to any pipe, ditch, channel, tunnel, conduit, discrete fissure, container, rolling stock, concentrated animal feeding operation, or vessel or other floating craft, from which pollutants are or may be discharged. (33 U.S.C. § 1362(14)).

Your inquiry must remain focused on the manner by which the stormwater is directed to waters of the United States. The distinction is often elusive, but should be carefully considered and weighed in light of the facts regarding the particular facility. One court has described the point source requirement in the following way.

> Simple erosion over the material surface, resulting in the discharge of water and other materials into navigable waters, does not constitute a point source discharge, absent some effort to change the surface, to direct the water flow or otherwise impede its progress. . . . Gravity flow, resulting in a discharge into a navigable body of water, may be part of a point source discharge if the [discharger] at least initially collected or channelled the water and other materials. A point source of pollution may also be present where [dischargers] design spoiled piles from discarded overburden such that, during period of precipitation, erosion of spoil pile walls results in discharges into a navigable body of water by means

of ditches, gullies and similar conveyances, even if
the [dischargers] have done nothing beyond the mere
collection of rock and other materials. . . . Nothing in
the Act relieves [dischargers] from liability simply
because the operators did not actually construct those
conveyances, so long as they are reasonably likely to
be the means by which pollutants are ultimately
deposited into a navigable body of water.
Conveyances of pollution formed either as a result of
natural erosion or by material means, and which
constitute a component of a . . . drainage system, may
fit the statutory definition and thereby subject the
operators to liability under the Act. *Sierra Club v.
Abston Construction Co., Inc.*, 620 F.2d 41, 45 (5th Cir.
1980).

In the preamble to the stormwater permit application
regulations, EPA commented on this case, stating:

Under this approach, point source discharges of
stormwater result from structures which increase the
imperviousness of the ground which acts to collect
runoff being conveyed along the resulting drainage of
grading patterns. (55 *FR*).

Thus, sheet flow off an industrial facility would generally
be included in the definition of "point source." However,
stormwater discharges comprised solely of natural sheet
flow from undeveloped areas located at an industrial
facility may not meet the point source requirement.

Second, the discharge must be to "waters of the United
States." EPA's regulations define "waters of the United
States" to include:

1.   all waters that are currently used, were used in the
past, or may be susceptible to use in interstate or foreign

commerce, including all waters which are subject to the ebb and flow of the tide

2. all interstate waters, including interstate "wetlands"

3. all other waters such as intrastate lakes, rives, streams (including intermittent streams), mudflats, sandflats, "wetlands," sloughs, prairie potholes, wet meadows, play lakes, or natural ponds the use, degradation, or destruction of which would affect or could affect interstate or foreign commerce including any such waters

(a) that are or could be used by interstate or foreign travelers for recreational or other purposes;

(b) from which fish or shellfish are or could be taken and sold in interstate or foreign commerce; or

(c) that are used or could be used for industrial purposes by industries in interstate commerce;

4. all impoundments of waters otherwise defined as waters of the United States under this definition

5. tributaries of waters identified in paragraphs (1) through (4) of this definition

6. the territorial sea and

7. "wetlands" adjacent to water (other than waters that are themselves wetlands) identified in paragraphs (1) through (6) of this definition

This definition does not include discharges to groundwaters, although recent court cases have indicated that EPA may have jurisdiction under the CWA over some discharges to groundwater connected to surface water. (*Inland Steel Co. v. EPA*, 901 F.2d 1419 (7th Cir. 1991)). EPA stated that the stormwater program does not cover discharges to groundwater "unless there is a hydrological connection between the ground water and a nearby surface water body." (55 *FR*). In many states the distinction is academic. States delegated to administer the CWA often substitute their own definition of "waters of the State" for EPA's definition of "waters of the United States" and include groundwater within their definition of "waters of the State." In these states there is no statutory exclusion for stormwater discharges to groundwater.

EPA's regulations also establish different permitting requirements for stormwater discharges to different types of sewers. Not all discharges to all types of sewer systems must obtain an NPDES stormwater permit, although they may still be regulated discharges under other permitting programs. First, the definition of "stormwater associated with industrial activity" specifically excludes discharges from facilities or activities that are excluded from the NPDES program under 40 CFR Part 122 (the indirect discharger exclusion). Consequently, stormwater discharges into Publicly Owned Treatment Works (POTWs) or a sanitary sewer system are not covered under the NPDES program, including the stormwater regulations. However, these discharges will usually be regulated under the POTW's pretreatment program.* Similarly, conveyances that discharge stormwater runoff combined with municipal sewage (*e.g.*, discharges *from* POTWs that have combined sewer systems) are point sources that must be authorized by NPDES permits in accordance with the procedures of 40 CFR § 122.21 but are not subject to the stormwater regulations because the discharges are not composed entirely of stormwater.

The rule does not except stormwater discharges associated with industrial activity that discharge through a "mu-

---

*See *U.S. EPA 1993, NPDES Storm Water Program Question and Answer Document: Vol. 2*, U.S. EPA, Office of Water, Washington, DC, EPA 833-F-93-0028 at p. 4.

nicipal separate storm sewer" from the stormwater permit requirement because these discharges do not go to a POTW (55 *FR*).[†] EPA decided that these types of stormwater discharges associated with industrial activity should be stormwater directly regulated through NPDES permits separate from those issued to the municipality. EPA nevertheless expects that most permits covering individual stormwater discharges associated with industrial activity that discharge to municipal separate storm sewers will require industrial dischargers also to comply with the terms of the NPDES permit issued to the municipality for the municipal separate storm sewer system, as well as the terms of the permittee's own NPDES stormwater permit.

All operators of industrial stormwater discharges that discharge into a nonmunicipal or nonpublicly-owned (*e.g.*, federally or privately owned) stormwater conveyance must obtain a stormwater permit for the discharge. This provision could be applicable to sewer systems at an industrial park with multiple tenants that ultimately discharges directly into navigable waters. These types of discharges must either be covered by an individual permit for each discharge to the nonmunicipal system or a single permit issued to the operator of the last portion of the system that discharges to waters of the United States, with each discharger to the nonmunicipal conveyance a co-permittee to that permit (40 CFR). The regulations define "co-permittee" as a "permittee to an NPDES permit that is only responsible for permit conditions relating to the discharge for which it is the operator." This provision provides protection to the last discharger before the wa-

---

[†]A municipal separate storm sewer is defined as "a conveyance or system of conveyances (including roads with drainage systems, municipal streets, catch basins, curbs, gutters, ditches, manmade channels, or storm drains) that is owned by a state or local government body, is designed for collection or conveying stormwater, is not a combined sewer and is not part of a POTW" (40 CFR § 122.26(b)(8)).

ters of the United States from violations caused by facilities discharging into the upper portions of the nonmunicipal system. Where there is more than one operator of a single system of nonmunicipal stormwater conveyances, all operators of industrial stormwater discharges must submit permit applications. Any permit that covers more than one operator must identify the effluent limitations, or other permit conditions, if any, that apply to each operator.

Stormwater Discharges Associated with Industrial Activity

If an industrial stormwater discharge is covered under the CWA, the next broad question to be addressed is whether the discharge is a "stormwater discharge associated with industrial activity." Dischargers of stormwater "associated with industrial activity" must apply for a stormwater permit.

The definition of "stormwater discharges associated with industrial activity" is complex and should be scrutinized carefully. "Stormwater" is defined as "stormwater runoff, snow melt runoff, and surface runoff and drainage" (40 CFR). This definition does not include infiltration or street wash waters (although these and similar discharges can be covered under EPA's industrial stormwater general permit, as discussed later). However, the definition does include discharges from retention or detention basins used to collect stormwater that are part of a conveyance system for stormwater associated with industrial activity and that discharge to waters of the United States.

"Stormwater discharges associated with industrial activity" is defined as stormwater discharges from one of the eleven categories of industrial activity specified in 40 CFR. The list includes virtually every significant type of manufacturing activity, and intends to exclude only commercial and retail activities. A facility must be engaged in the listed types of industrial activity before it is subject to the stormwater regulations. Five of these industrial activity categories are defined by Standard Industrial Classification (SIC) codes, and six are defined through narrative descriptions of the covered industrial activity.

In its November 1990 permit application rule, EPA divided the covered industrial categories into two groups. One group consists of the first ten categories, generally referred to as "heavy" industry. Facilities in these ten industrial categories must obtain a stormwater permit for stormwater discharges associated with industrial activity whether or not the industrial activities are actually exposed to stormwater. The second group, often called "light" industry, consists of a very large eleventh category. Facilities in this eleventh category need to apply for a stormwater permit only if stormwater physically contacts products, materials, materials handling equipment or activities or other industrial activity equipment or sites (the "no exposure" exemption). The distinction between the two groups was based on the premise that more "light" industrial activity occurs indoors, and these types of facilities should have an opportunity to keep work areas and materials away from contact with stormwater and avoid regulation entirely. The definition of what constitutes stormwater discharges associated with industrial activity is also slightly different for the two groups.

However, the United States Court of Appeals for the Ninth Circuit struck down EPA's distinction between light and heavy industry in *NRDC v. EPA*. 966 F.2d 1292 (9th Cir. 1992). The court concluded that the WQA phrase "associated with industrial activity" was very broad, and reflected Congress' intent to exclude only discharges from nonindustrial parts of facilities, such as parking lots. The court vacated the "light" industry "no exposure" distinction and remanded the rule to EPA for further proceedings. On December 18, 1992, EPA issued a rule stating its position on the impact of the court remand (57 *FR.*) In the rule EPA stated that it will retain the "light" industry distinction until it conducts a further rulemaking to address the court's concerns. EPA also stated it will revisit the light industry distinction at that time and determine whether to reissue the rule. EPA's position is that facilities attempting to exempt themselves from stormwater permit application requirements under the light industry exemption should await further rulemaking action before attempting to seek stormwater permits.

The first group of ten "heavy" industrial categories includes:

1.   facilities subject to national effluent limitations guidelines under 40 CFR subchapter N

2.   facilities classified as Standard Industrial Classifications 24, 26, 28, 29, 311, 32, 33, 3441, and 373 with a few exceptions for certain industries within the major categories

3.   facilities classified as Standard Industrial Classifications 10 through 14

4.   hazardous waste treatment storage and disposal facilities

5.   landfills, land application sites, and open dumps that receive or have received industrial wastes. Industrial wastes are defined as any materials that originated from any of the types of facilities that are included under the definition of "industrial activity" (55 FR)

6.   facilities involved in the recycling of materials, including metal scrap yards, battery reclaimers, salvage yards and automobile junk yards including, but not limited to, those classified as Standard Industrial Classifications 5015 and 5093

7.   steam electric power generating facilities, including coal-handling sites

8.   transportation facilities classified as Standard Industrial Classifications 40–45 (with some exceptions) that have vehicle maintenance shops, equipment cleaning operations, or airport de-icing operations

9.   treatment works treating domestic sewage

10.   construction activities, including clearing, grading, and excavation activities that result in the disturbance of five acres total land or more

The tenth category, listed above, includes construction activities only if the activity disturbs five acres total land or more. In *NRDC v. EPA*, 966 F.2d 1292 (9th Cir. 1992) the court also remanded EPA's regulations exempting construction activities involving less than five acres. As in the case of the "light" industry exemption remand, EPA has stated it will retain the less-than-five-acre exemption until it conducts further rulemaking activities (57 *FR*).

The regulations also establish an eleventh group of "light" industrial facilities, classified as Standard Industrial Classifications 20–23, 2434, 25, 265, 267, 27, 283, 285, 30, 31, 323, 34–39, and 4221–25 (with a few exceptions), that are not otherwise included in heavy industry categories 2–10. Stormwater discharges from these facilities need to be permitted only if certain materials or activities are *exposed* to stormwater (40 CFR).

Under this "light industry no exposure" exemption, a facility that conducts all specified "industrial activities" inside or in an otherwise covered manner does not need to apply for a stormwater permit. EPA has not clarified to what extent industrial activities must be protected to qualify for the "no exposure" exemption available to facilities in category 11. The preamble notes only that the exposure exemption is appropriate for "wholly enclosed" facilities or facilities that have "their operations entirely protected from the elements" (55 *FR*). However, recent EPA guidance documents suggest that EPA will adopt, on a case-by-case basis, a "reasonable potential" standard for exposure. State permitting authorities may adopt a different, more stringent, standard, however.

The list of industrial activities or applicable SICs will define what facilities must apply for a stormwater permit. Facilities that warehouse finished products under the same SIC code at a different facility from the site of manufacturing do not need to file a permit application, unless otherwise covered by the rule (55 *FR*). The regulatory status of off-site facilities must be independently examined. First, off-site facilities may be classifiable according to their own SIC code. If there is no SIC code that describes the off-site facility independently, then it would assume the SIC code of the parent facility it supports. However, certain off-site facilities classified as SIC code auxiliary facilities or included within the facilities' primary SIC code description would usually be classified according to the parent facility they support. Such supporting establishments include central administration offices, research and development laboratories, maintenance garages, and local trucking terminals.

Facilities with multiple functions (such as industrial and retail activities) need only submit a permit application for the covered industrial portion of the facility, so long as the stormwater from the nonindustrial portion is segregated. Discharges from undeveloped areas and areas not encompassed by the definitions of industrial activity, such as parking lots, and administrative or employee buildings do not need to be permitted. A facility can also choose to segregate its discharges into those associated with industrial activity and those not associated with industrial activity, or into "light" and "heavy" industry discharges.

EPA does, however, have the authority under Section 402(p)(2)(E) of the CWA to require a permit for stormwater discharges excluded from the definition of "associated with industrial activity," by designating stormwater discharges, such as those from large parking lots, that are significant contributors of pollutants or contribute to a water quality standard violation. Municipalities engaged in any of the covered industrial activities are covered by the stormwater program in the same manner as industrial facilities. However, the Intermodal Surface Transportation Efficiency Act of 1991 exempts, for now, municipalities with a population of less than 100,000 from the industrial activity permit requirement, except for airports, power plants, or uncontrolled sanitary landfills (57 *FR*).

The regulations define stormwater "associated" with industrial activity for the first ten "heavy" industry categories as

> the discharge from any conveyance which is used for collecting and conveying stormwater *and which is directly related to manufacturing, processing or raw materials storage areas at an industrial plant* (40 CFR).

This broad definition should be used to resolve uncertainties. EPA's regulations further elaborate on this definition, primarily by way of example. The definition specifically includes language to the effect that it is not a comprehensive list, but only illustrative. The list

> includes, *but is not limited to*, plant stormwater discharges from industrial yards; immediate access roads and rail lines used or traveled by carriers of raw materials, manufactured products, waste material, or by-products used or created by the facility; material handling sites; refuse sites; sites used for the application or disposal of process waste

waters; sites used for the storage and maintenance of material handling equipment; sites used for residual treatment, storage or disposal; shipping and receiving areas; manufacturing buildings; storage areas (including tank farms) for raw materials, and intermediate and finished products; and areas where industrial activity has taken place in the past and significant materials remain and are exposed to stormwater (40 CFR)

Note that the term "significant materials" as used above in relation to past industrial activities is defined as follows:

Significant materials includes, but is not limited to: raw materials; fuels; materials such as solvents, detergents, and plastic pellets; finished materials such as metallic products; raw materials used in food processing or production; hazardous substances designated under section 101(14) of CERCLA; any chemical the facility is required to report pursuant to section 313 of Title III of SARA; fertilizers; pesticides; and waste products such as ashes, slag and sludge that have the potential to be released with stormwater discharges.

The regulations also define the term "material handling activities" as including: the storage, loading and unloading, transportation, or conveyance of any raw material, intermediate product, finished product, by-product, or waste product.

The definition of "associated with industrial activity" focuses on specific areas and areas where specific activities occur. The first sentence establishes the broad scope of the definition by covering "discharges from plant yards." The rest of the definition simply adds, by illustration, additional plant and non-plant yard sites such as immediate access roads, which include only "roads which are exclusively or primarily dedicated for use by the industrial facility" and which do not include state, county, or federal roads (55 *FR*); waste disposal sites, storage sites, shipping and receiving areas, manufacturing buildings, and past activity areas. For facilities in the ten "heavy" industry categories, stormwater discharges associated with the activities described above must be permitted, regardless of the actual exposure of the activities to stormwater, and even if the activities or materials (such as storage facilities) are enclosed (55 *FR*).

For facilities in the eleventh, "light" industry category, the definition of "stormwater discharges associated with industrial activity" is slightly different. The definition slightly modifies the "heavy" industry requirements. It is intended to exclude stormwater discharges at light industry facilities where the stormwater is not contaminated through contact with industrial activities or areas. The regulations state:

> For the categories of industries identified in paragraph (b)(14)(xi) (*i.e.*, "light industry") of this section, the term includes only stormwater discharges from all the areas (except access roads and rail lines) that are listed in the previous sentence (relating to "heavy" industry) where material handling equipment or activities, raw materials, intermediate products, final products, waste materials, by-products, or industrial machinery are *exposed* to stormwater.

There are some subtle differences in this light industry "exposure" definition of "associated with industrial activity." First, discharges from access roads and rail lines are excluded. Second, a two-step process must be met for the "light" industries. First, the discharge must come from an "area" described in the heavy "associated with industrial activity" definition. Next, material handling equipment or activities (a defined term), raw materials, intermediate products, final products, waste materials, by-products, or industrial machinery *located in those areas* must be exposed to stormwater.

In the preamble to the permit application regulations, EPA elaborated on certain types of activities or facilities that, in its view, are considered to be "associated with industrial activity." EPA included stormwater discharges from drainage ponds. EPA's regulations state that the term "storage areas" is "associated with industrial activity" and includes "tank farms." In the permit application rule preamble, EPA stated that tank farms are in existence to store products and materials "created or used" by the facility, and are therefore directly related to the manufacturing process. This interpretation would include the storage of fuel oil. (The definition of "significant materials" also specifically includes the term "fuel," although the regulations use only the term "significant materials" in connection with past activities.)

Industrial Stormwater Permit Applications

EPA's industrial stormwater permit application rule, issued November 16, 1990, 55 *FR* 47990, requires all covered industries to file an NPDES permit application for all stormwater discharges "associated with industrial activity." This rule established three types of possible permit applications that would result in two possible types of permits. Although the types of available permit applications and permits is relatively straightforward, the issue becomes somewhat confused under statutes like the CWA, which allows delegation of program administration to the states.

Types of Permits

EPA is authorized to issue two types of NPDES permits: individual permits and general permits. Individual permits, issued to individual facilities on a case-by-case basis with terms and conditions appropriate to the particular facility, are used almost exclusively in the traditional NPDES process wastewater permitting program. EPA's pre-existing regulations also authorized use of general permits, although they were not often used in the NPDES permit program (40 CFR). General permits cover a category of discharges or facilities within specified geographic areas. General permit terms and conditions are generally similar to those in individual permits, except that general permits can authorize discharges from more than one facility. General permit terms and conditions are determined by notice and comment rulemaking. This is the only opportunity for dischargers to comment on general permits. EPA regulations provide that a discharger may always request an individualized permit determination. Once a discharger agrees to sign on to a general permit, usually by notifying the permitting agency in writing of acceptance of the general permit, the general permit and all its terms and conditions apply to the discharger just as if the permit were an individually issued permit. In exchange for permitting convenience, a discharger accepting a general permit can negotiate the terms and conditions of the general permit only through rulemaking comments, and cannot appeal the terms and conditions of the permit before the permitting agency.

EPA does not administer the NPDES program in most states and territories. CWA Section 402(b) allows any state to petition EPA for approval to administer the CWA, and many states have done so (33 U.S.C.). In order to issue general permits, a state must have in its program EPA-approved general permitting authority. Without this authority, a state is not authorized to issue general NPDES permits, and may issue only individual permits. At the time that EPA issued the stormwater regulations, many states did not have EPA-approved general permitting programs, leading to uncertainty as to whether general permits would be available in many states. However, as of late 1993, most delegated states had obtained general permitting authority. The following summary shows the breakdown on permitting authority.

- 40 states and territories are delegated to administer EPA's NPDES program, and issue NPDES permits in those states.
- Of these 40 states and territories, only three (Kansas, Michigan, and the Virgin Islands) do not have general permitting authority. In those three states general permits for stormwater discharges are not available. In the other 37 delegated states, states may issue, by rule, general permits for stormwater discharges. A few states apparently may choose not to issue general permits for industrial stormwater discharges at all.
- There are 16 states and territories that are not delegated to administer the NPDES program. In these states and territories, EPA issues NPDES permits directly. EPA's general permits are available in all these states and territories. The CWA authorizes the EPA Administrator to issue NPDES permits, but this authority has been delegated to EPA's ten Regional Administrators (40 CFR).

Types of Applications

EPA's permit application rule also authorized three methods to apply for permit coverage:

1. individual permit applications

2.   group permit applications (may result in either an individual or general permit)

3.   general permit applications (notifications of intent to be covered or NOIs).

Under EPA's regulations, an individual stormwater permit application for an existing facility was to be filed by October 1, 1992, unless the discharger is part of a group application or covered by a stormwater general permit. Many dischargers filed individual permit applications to meet the October 1, 1992 application deadline and then later withdrew these applications and opted for coverage under EPA or state-issued general permits, when these became available.

Dischargers filing individual permit applications are required to complete EPA's Form 2F (Application for Permit to Discharge Stormwater Associated with Industrial Activity) as well as EPA's Form I (general information). Most, but not all, delegated states use or accept EPA's permit application forms. The application requires an extensive amount of information, including significant monitoring data; a topographic map and site drainage plan; a narrative description of "significant materials" that in the past three years have been treated, stored, or disposed in a manner to allow exposure to stormwater; a certification that all outfalls that should contain stormwater associated with industrial activity have been tested or evaluated for the presence of non-stormwater discharges that are not covered by an NPDES permit; and existing information on "significant" leaks or spills of toxic or hazardous pollutants at the facility in the past three years. Individual permit application details are covered elsewhere in this book.

EPA's permit application regulations also authorize the use of group applications. Group applications are one of

EPA's primary strategies in issuing stormwater permits to existing dischargers. The group application process was designed as a one-time administrative procedure to manage the task of getting existing industrial dischargers covered by stromwater permits (57 *FR*). A group application may be filed by an entity representing a group of applicants that:

- are part of the same EPA national "effluent limitations guidelines" subcategory (see 40 CFR Subchapter N, parts 405 to 471 for listing of industrial categorical standards); or
- are sufficiently similar to be appropriate for general permit covered under the factors specified at 40 CFR § 122.28(a)(2); these factors include sources that involve the same type or similar operations, discharge the same type of wastes, require the same effluent limitations or operating conditions, or require the same or similar monitoring requirements.

There are two primary advantages to group applications. First, sampling requirements for application submission are substantially reduced. Second, the permit terms and conditions can be negotiated and tailored to fit more precisely the permitting needs of particular groups than would be the case under a general permit covering all types of industrial facilities.

Part 1 of a group permit application was due at EPA by September 30, 1991. Facilities not joining an initial group application were allowed to petition for inclusion in a group based upon a showing of good cause by February 18, 1992. However, as discussed later, EPA is proposing to allow any discharger that fits within specified industrial categories to opt for coverage under the permits issued to group applicants.

The Part 1 application requires only relatively general information about group members to allow EPA to determine whether each member is properly part of the group. Based on the information in Part 1, EPA is supposed to approve or deny the members of the proposed group within 60 days. Rejected group members have one year to file individual applications or obtain general permit coverage. Facilities that are owned or operated by municipalities and that are rejected from group applications have until the later of 180 days after the notice of rejection or October 1, 1992 to file an individual application.

Part 2 of the group application was to be submitted to EPA by October 1, 1992. Part 2 of the application consists primarily of the representative sampling data from the dischargers designated to supply these data. EPA may also require other information generally required under the Form 2F individual permit application (55 *FR*).

All group permit applications are to be submitted to EPA. EPA will then distribute summaries to NPDES-authorized states. Once group applications are reviewed, EPA will develop group permit terms and conditions and issue permits covering group applications located in nonauthorized states. NPDES-authorized states can then adopt, with or without modifications, EPA's group permits for facilities in their states if they choose.

A group application can result in either a general permit or an individual permit. In states where EPA is the permit-issuing authority, EPA will issue a general permit applicable to all the members of a particular group. However, prior to issuance of an EPA NPDES permit in nondelegated states, the state must issue a CWA Section 401 water quality certification (33 U.S.C.). Through these certifications states can impose additional permit requirements into EPA's NPDES permits, including requirements to ensure that the dischargers authorized will ensure compliance with all applicable water quality standards.

In the case of NPDES-delegated states, each state must decide what type of permit to issue. If the state has general permitting authority, the state can issue a general permit covering all group members within the state. If the state does not have general permitting authority, it can still use the EPA model group permits as the basis for individual permits issued individually to group members within the state. However, delegated states are under no obligation to make use of EPA's group permits at all or in the form approved by EPA. A few states have indicated that they do not plan to make use of EPA's final group permits when issued, and others have indicated they are not sure what they will do. A state may also adopt parts of EPA's model group permits and reject or modify others. In any case, EPA must issue its final group permits first. Then the states must draft or otherwise adopt these permits. The permits then must go through a public comment period at the state level (either as a rulemaking in the case of general permits or public notice in the case of individual permits). In either case, it will take some time after EPA issues its final model group permits before they can be in place in NPDES-delegated states.

Finally, in its permit application rule EPA declared its intent to issue one or more general permits (55 *FR*). Dischargers can then file applications to be covered under a general permit. Dischargers covered by a general permit are excluded from the requirement to submit individual or group permit applications. On April, 1992, EPA published a final rule establishing final minimum requirements for filing an application to be covered by a general permit (57 *FR*). These requirements generally apply as well to state-issued general permits. General permit applications are typically called "Notifications of Intent" (NOIs). NOI requirements established for general permits operate instead of individual permit application requirements. NOI requirements are to be specified in the EPA or state general permit (40 CFR). At a minimum these must include the name and address of the owner or operator, the facility name and address, the types of facilities and discharges to be covered and the names of the receiving waters. NOIs must be signed in accordance with the signatory requirements of 40 CFR Part 122.22. General permits may require that additional information be reported in NOIs. The general permit will also specify deadlines for submitting NOIs and the date when a discharger is authorized to discharge under the permit.

The general permit itself will describe the types of facilities eligible for coverage under the permit. General permit coverage provides an easy method of obtaining a stormwater permit, because no sampling is generally required under the NOI provisions and because the permit is usually effective within a few days of submission of the NOI.

Industrial Stormwater Permit Terms and Conditions

Introduction

The 1987 WQA clarified that individual stormwater permits must include all applicable provisions of CWA Sections 301 and 402. Industrial stormwater permits must address the same statutory requirements as NPDES process wastewater permits. Permits must include technology-based controls based on Best Available Technology (BAT) or Best Control Technology (BCT) or, if necessary, water-quality-based standards. Typically, EPA implements the BAT/BCT requirement by developing numeric end-of-pipe effluent limitations based upon the performance of selected technology, effectively requiring dischargers to employ the selected treatment technology to meet the numeric effluent limitations. In order to ensure compliance with applicable numeric permit effluent limits, dischargers are also required to sample their effluent periodically, and report on their compliance status.

As a result of EPA's promulgation of two "core" general permits, one for stormwater discharges associated with construction sites and one for stormwater discharges associated with industrial activity—as well as other EPA stormwater permit rulemakings—the outlines of the permit terms and conditions that EPA will require under its stormwater program are available. These rules establish EPA's minimum permit requirements (i.e., EPA's initial determination of what practices and controls constitute BAT/BCT based on Best Professional Judgment [BPJ]). They should form the basis for any individual stormwater permits issued by EPA. In addition, they should form the general outline of state general and individual stormwater permits, although the states will also have to add any water-quality-based practices and controls that are necessary to meet state water quality standards. In addition, different BAT/BPJ standards may be appropriate for particular facilities. Under the CWA, state stormwater permits need only be as stringent as EPA's permits, but can be more stringent if the state so chooses. States have generally closely followed EPA's core general permits; however, there are differences in particular requirements from state to state.

General Permit—Stormwater Pollution Prevention Plans

In the core general permits for construction sites and other industrial activity, EPA took a very different approach to implementing the CWA's BAT/BCT requirements (57 *FR*). EPA determined in almost all cases that site-specific stormwater pollution prevention plans (SPPPs) to be developed and implemented to minimize and control pollutants in stormwater discharges were sufficient to achieve the CWA's BAT/BCT requirements, in lieu of numeric effluent limitations. The only numeric effluent limitation established in EPA's core general permit for industrial activity is for total suspended solids and pH applicable to discharges of coal pile runoff. There are no numeric effluent limitations in the general permit for construction sites. However, EPA's industrial general permits also include state-specific conditions based on water quality standards that may include additional numeric effluent limits.

The heart of EPA's industrial general permit is the facility SPPP. All facilities covered by EPA's general permits must develop, submit, and implement an SPPP. Most existing dischargers must develop an SPPP by April 1, 1993, and implement the plan by October 1, 1993. SPPPs for facilities where industrial activity commences after October 1, 1992, but on or before December 31, 1992, must be developed by and provide for compliance with the plan within 60 days of commencement of industrial activities. SPPPs for facilities commencing operations on or after January 1, 1993 must be developed and provide for compliance on or before the data of submission of the NOI to be covered under the general permit. The industrial general permit specifies slightly different deadlines for discharges from oil and gas exploration, production, processing, or treatment operation or transmission facilities, and for facilities owned or operated by a municipality that has participated in a timely group application where the municipality's application is rejected or the facility is denied participation in the application. EPA provided additional time to comply with certain special SPPP requirements applicable to EPCRA section 313 facilities and salt storage facilities.

SPPPs must identify sources of pollution potentially affecting the quality of stormwater discharges, describe and ensure implementation of practices or measures to minimize and control pollutants in stormwater from the facility, and ensure compliance with the terms and conditions of the permit. All facilities must meet the following minimum SPPP requirements.

1. *Pollution Prevention Team.* The permittee must identify a qualified individual or team to be responsible for developing and implementing the plan.

2. *Description of Potential Pollution Sources.* SPPPs must describe site activities, materials, and physical features that may contribute significant amounts of pollutants to stormwater, or, during dry weather, result in pollutant discharges through the separate storm sewers or stormwater drainage systems that drain the facility. The following elements must be specifically described:

- an inventory of exposed significant materials
- a list of significant spills or leaks of toxic or hazardous pollutants
- a certification that discharges from the site have been tested for the presence of non-stormwater discharges
- a description of any existing stormwater sampling data
- a narrative risk assessment and summary of potential pollution sources.

3. *Measures and Controls.* The SPPP must evaluate, select, and describe the pollution prevention measures, best management practices (BMPs) and other controls that will be implemented at the facility. The SPPP must contain a schedule for implementing each control or practice selected. The discussion of measures and controls must address the following minimum components:

- good housekeeping practices, protocols, and techniques
- a preventative maintenance program for stormwater management devices and other equipment and systems
- spill prevention and response procedures
- site inspections
- employee training
- recordkeeping and internal reporting procedures
- sediment and erosion control for areas with a high potential for significant soil erosion
- evaluation of stormwater management practices that divert, infiltrate, reuse, or otherwise manage stormwater runoff to reduce the discharge of pollutants.

4. *Comprehensive Site Evaluation.* The SPPP must describe the scope, content, and frequency of comprehensive site inspections that (1) confirm the accuracy of the plan's description of potential pollution sources; (2) determine the effectiveness of the plan; and (3) assess compliance with the terms and conditions of the permit.

EPA's inclusion in the plan requirements of a schedule of compliance for implementing BMPs and other measures and controls suggests that a facility does not need to complete, for example, construction of facilities by the October 1, 1993 date for plan implementation, so long as the facility sets out a reasonable compliance schedule in the plan and adheres to it.

The industrial general permit contains additional SPPP requirements for facilities subject to the annual reporting requirements of EPCRA Section 313 for water priority chemicals. EPCRA reporting requirements are provided in 40 CFR Part 372.26. Further, a list of the 175 EPCRA

water priority chemicals are provided in 57 *FR* 41246 and
41331-35. In addition to the SPPP requirements listed
above, these facilities must include in their SPPPs special
provisions that address water priority chemical storage,
processing, or handling areas, and must

- provide for appropriate containment, drainage
  control, and/or diversionary structures for water
  priority chemical storage, handling, or processing
  areas
- implement appropriate measures to address the
  following types of facility priority areas:
  - liquid storage areas where stormwater comes into
    contact with any equipment, tank, container, or
    other vessel used for Section 313 water priority
    chemicals
  - material storage areas for Section 313 water prior-
    ity chemicals other than liquids
  - tank and rail car loading and unloading areas for
    liquid Section 313 water priority chemicals
  - areas where Section 313 water priority chemicals
    are transferred, processed, or otherwise handled
- address preventative maintenance and
  housekeeping
- address facility security
- address facility training
- provide a review and certification of the special
  requirements by a Registered Professional Engineer.

Facilities that had to report releases under EPCRA Section
313, but that, during the term of the permit, no longer
manufacture, process, or otherwise use EPCRA Section
313 water priority chemicals on-site in amounts that exceed the applicable EPCRA 313 thresholds, are not subject
to the special EPCRA Section 313 requirements after the
reductions are made. The additional requirements for

EPCRA Section 313 facilities and salt storage facilities do not have to be met until October 1, 1995 (57 *FR* 41308). In addition, a facility that triggers EPCRA 313 thresholds for the first time during the permit term has three years from the first time it has to report to meet the special requirements for these facilities.

SPPPs must be kept current. Plans are to be amended whenever there is a change in design, construction, operation, or maintenance that "has a significant effect on the potential for the discharge of pollutants" or if the plan proves to be ineffective in eliminating or significantly minimizing pollutants. The SPPP as well as all NOIs, reports, certifications, or information must be signed in accordance with the particular specifications set out in Part VII. G. of the permit. The permittee must retain the plan on-site until at least one year after coverage under the permit ends. Records of all monitoring information, reports, and records of data used to complete the NOI must be kept for the same period of time, except for monitoring data required under the permit, which must be kept for six years from the sampling date. The general permit expires on October 1, 1997, but continues in full force and effect until a new permit is issued.

The industrial general permit contains a number of other provisions that must be complied with. Operators of stormwater discharge through "large" or "medium" municipal separate storm sewer systems must send a copy of their NOI to the municipal operator. The general permit does not authorize non-stormwater discharges. However, a number of non-stormwater discharges can be authorized under the permit if identified in the SPPP. These non-stormwater discharges include: firefighting discharges and fire hydrant flushings, potable water sources, irrigation drainage, lawn watering, routine external building washdowns without detergents, pavement washwaters for areas where spills of hazardous or toxic materials have not occurred or have been cleaned up, air conditioner condensate, uncontaminated groundwater, and other listed discharges. The general permit requires that releases of hazardous substances or oil from a facility must be eliminated or minimized in accordance with the SPPP. When stormwater contains oil or hazardous substances in amounts exceeding the reportable quantities established under 40 CFR 110, 40 CFR 117, or 40 CFR 302, appropriate notifications must be made, the SPPP must be modified within 14 days of the release, and a report on the release must be submitted to EPA, also within 14 days.

Under Section 401 of the CWA, nondelegated NPDES states must certify to EPA that any NPDES permit will comply with applicable state water quality standards under the CWA. A number of nondelegated states added a number of specific provisions to EPA's general permit as part of their Section 401 certification. The specific general permit for each state should be carefully reviewed to identify these additional conditions. For example, Maine requires use of different organisms for WET testing, while New Mexico added a number of numeric water-quality-based "action levels" for stormwater discharges into certain domestic water supply water bodies. Texas also added a number of numeric water-quality-based effluent limits for discharges into certain water bodies (57 FR).

Group Permits—EPA's Proposed Multi-Sector Permit
On November 19, 1993, EPA published its proposed Stormwater Multi-Sector Permit, in response to most of the group permit applications it received (58 FR). EPA initially received over 1,200 group applications, but ultimately combined them into 29 broad "industrial sectors," based on specified similarities. Under the proposal, coverage under the multi-sector permit is available to group applicants, and to any other new or existing facility that is part of one of the 29 covered industrial sectors, including

dischargers already covered under EPA's baseline industrial general permit. Like EPA's baseline industrial general permit, the proposed multi-sector permit will cover only facilities in the 12 states and territories where EPA administers the NPDES program. NPDES-delegated states will have to decide whether they will adopt the multi-sector permits when they are final. About a dozen NPDES-delegated states have indicated that they do not plan to use EPA's multi-sector permit in their states. Most NPDES-delegated states plan to issue similar multi-sector permits in their states, based, at least in part, on EPA's final multi-sector permit.

There are a number of major proposed changes in the multi-sector permit that are different from EPA's baseline industrial general permit. Facilities that can qualify under either permit will need to evaluate carefully the pros and cons of each permit. Permittees should note that states may not adopt final multi-sector permits based on EPA's permit until well into 1995. However, some states are requiring permits now. Obligations under these deadlines should be reviewed prior to selecting a permit coverage option.

Some of the more significant changes in the proposed multi-sector permit include the following.

- There are no sampling requirements for all EPCRA Section 313 facilities, although 17 our of 29 industry sectors must still sample.
- Sampling is required only in years two and four of the five-year permit; not semiannually or annually.
- If a facility's initial monitoring program establishes that certain stormwater pollutant parameters are discharged below specified "benchmark" values (based primarily on EPA's very stringent "Gold Book" water quality criteria), the facility would not need to do further sampling during its permit term.

- The multi-sector permit is also based around SPPPs. Although the multi-sector permit SPPP requirements are similar in form to the baseline industrial general permit SPPP requirements, there are additional detailed industry-specific requirements for most industrial sectors.
- The only numeric effluent limitations in the proposed multi-sector permit are from existing categorical effluent limitation guidelines for the asphalt, emulsion manufacturing, and allied chemicals manufacturing industries. Again, some states can be expected to add additional numeric effluent limitations based on state water quality standards.

About a dozen NPDES-delegated states have indicated they do not plan to use EPA's multi-sector permit in their states. Most NPDES-delegated states plan to employ similar multi-sector permits based at least in part on EPA's permit.

Stormwater permits must also contain monitoring requirements. In a final rule issued April 2, 1992 in 57 *FR* 11403, EPA established final minimum monitoring requirements for stormwater permits. The final rule provides for establishing monitoring conditions in stormwater industrial permits for discharges associated with industrial activity on a case-by-case basis. However, at a minimum, where a permit does not contain end-of-pipe numeric effluent limitations, but requires only a SPPP, the permit must require a discharger "to conduct an annual inspection of the facility site to identify areas contributing to stormwater discharges associated with industrial activity and evaluate whether measures to reduce pollutant loadings identified in a stormwater pollution prevention plan are adequate and properly implemented in accordance with the terms of the permit, or whether additional controls are needed" (40 CFR). A record summarizing the results of the inspection and a certification that the facility is in compliance with the permit, as well as identification of any incidents of noncompliance must be maintained for three years. The report and certification are subject to the signatory requirements of 40 CFR Part 122.22.

The requirements for SPPP site inspections in EPA's final core general permit for industrial facilities are consistent with EPA's April 2, 1992 final rule. The industrial general permit requires annual site inspections. Comprehensive site compliance evaluations are to be developed and must evaluate the facility's compliance with its SPPP and with the other terms and conditions of the general permit (57 *FR*). The general permit itself provides a detailed description of the elements required for a site compliance evaluation. The general permit requires that the evaluation be conducted by "qualified personnel." EPA's elaboration on what constitutes qualified personnel can be found at 57 *FR* 41275.

EPA's industrial general permit also establishes specific monitoring requirements for certain targeted industrial categories that are likely sources of significant amounts of pollutants. The specific monitoring requirements vary by the type of industrial facility involved. EPA established monitoring requirements to help identify pollutant sources, evaluate the effectiveness of pollution prevention plans, and evaluate the risk of discharges and potential water quality impacts. Because stormwater discharges are subject to water quality standards, if a facility's monitoring data indicate a reasonable potential to impact water quality, numeric end-of-pipe effluent limits may be added to the permit, although it is likely EPA and the states would await full implementation of SPPPs before taking such action. Permittees should know that EPA's industrial general permit contains a reopener clause for "potential or realized water quality impacts" that allows EPA to add additional requirements, including, possibly, end-of-pipe numeric effluent limitation to a facility's permit, if appropriate.

Facilities that are subject to SARA Title III, Section 313 of EPCRA for chemicals that are classified as "water priority chemicals" are required to perform semiannual monitoring of stormwater that is discharged from the facility and comes into contact with any equipment, tank, container, or other vessel or area used for storage of water priority chemicals, or located at a truck or rail car loading or unloading area, for nine specified parameters.

The following types of industries must conduct semiannual monitoring for the parameters specified for each industry group in the general permit:

- primary metals industries (SIC 33)
- land disposal units/incinerators/boilers and industrial furnaces
- wood treatment facilities
- coal pile runoff
- battery reclaimers

The following types of industries must conduct annual monitoring for the parameters specified for each industry group in the general permit:

- airports
- coal-fired steam electric facilities
- animal handling/meat packing
- additional miscellaneous facilities identified in 57 *FR* 41313

The general permit specifies two six-month sampling periods for each type of facility, and specifies when the sampling reports (Discharge Monitoring Report or DMR) must be submitted to EPA (57 *FR*). The general permit also specifies the sampling type requirements (grab or composite), appropriate storm events for sampling, and other monitoring requirements.

Many types of industries are required to conduct controversial acute whole effluent toxicity (WET) tests. EPA sampling requirements for such tests are detailed in 57 *FR* 41314. EPA considers WET testing to be an inexpensive method for evaluating the discharge of hundreds of particular toxic pollutants. A discharger subject to a WET testing requirement may, in lieu of monitoring for acute WET, monitor for all the pollutants listed in Tables II and III of Appendix D of 40 CFR 122, that the discharger "knows or has reason to believe are *present* at the facility site." Facilities subject to the acute WET monitoring requirement that detect toxicity on or after October 1, 1995 must review their SPPPs and make appropriate modifications to assist in identifying the sources of toxicity and to reduce toxicity in their stormwater discharges (57 *FR*).

There are some exceptions to the general permit monitoring requirements. First, when a facility has two or more outfalls that the permittee reasonably believes discharge substantially identical effluents, the permittee may monitor only one of the outfalls and report that the data also apply to the other substantially identical outfalls. In determining whether outfalls discharge substantially identical pollutants, the permittee must consider industrial activity, significant materials, and management practices and activities within the area drained by the outfall. The permittee must also include in its SPPP a description of the outfall locations and detail why the outfalls are expected to discharge substantially identical pollutants.

Second, a discharger may avoid the monitoring requirements altogether if the discharger makes an annual certification, under penalty of law, for a given outfall, that material handling equipment or activities, raw materials, intermediate products, final products, waste materials, by-products, industrial machinery or operations, significant materials from past industrial activity (or, for airports, de-icing activities) that are located in areas of the facility that are within the drainage area of the outfall are not presently exposed to stormwater and will not be exposed to stormwater for the certification period. The certification must be included in the SPPP and submitted to EPA.

Construction Site Permit Terms and Conditions

On September 9, 1992, EPA also issued a general permit covering stormwater discharges from construction sites as stated in 57 *FR* 41176. The construction site general permit is similar to the industrial general permit. The final general permit covers both existing construction sites (sites where construction began prior to October 1, 1992, and final stabilization occurs after October 1, 1992) and new construction sites (sites where disturbances associated with construction activities commence after October 1, 1992). To be covered under the permit a discharger must submit an NOI providing information similar to the industrial general permit NOI. NOIs for existing sites were due October 1, 1992, and NOIs for new sites are due two days prior to commencement of construction activities (*e.g.,* the initial disturbances of soils associated with clearing, grading excavation activities, or other construction activities).

Like the industrial general permit, the construction general permit also prohibits any non-stormwater discharges, except for certain specified types of discharges (*e.g.,* vehicle washwater or dust control water) that are specifically addressed in the SPPP. The construction site general permit also requires permittees to comply with the same requirements regarding releases in excess of reportable quantities of oil or hazardous substances as the general permit for industrial activities.

The heart of the construction site general permit is also

a stormwater pollution prevention plan. SPPPs for construction sites must be completed prior to submission of
the NOI. For construction activities that commenced on or
before October 1, 1992, the plan must provide for compliance with the terms and any schedule in the plan by
October 1, 1992, except that it may provide for compliance
with required sediment basins by no later than December
1, 1992. For construction activities that have begun after
October 1, 1992, the plan must provide for compliance
with the terms and any schedule in the plan beginning
with the initiation of construction activities.

Construction site SPPPs must include the following
elements:

1. *Site description.* The plan must contain a site
description that includes identification of potential
sources of pollution and a description of the site itself,
with site map and drainage patterns. The plan must
describe the specific construction project and activities
in detail, as specified in the general permit.

2. *Controls to reduce pollutants.* The SPPP must describe
and ensure implementation of appropriate measures
and controls. The plan must specifically address:

- erosion and sediment controls, including
  stabilization practices and structural practices
- stormwater management measures
- other specified controls (*e.g.*, to prevent discharges
  of building material waste or reduce dust
  generation)
- state and local sediment and erosion or stormwater
  management planning and control requirements.

3. *Maintenance.* Plans must contain a description of
maintenance and repair procedures addressing all
measures identified in the plan, including erosion and

sediment control measures and vegetation, to ensure such measures are kept in good operating condition.

4. *Inspections.* The plan must provide for inspections by qualified personnel of disturbed areas, areas used for storage of materials that are exposed to precipitation, structural control measures, and locations where vehicles enter or exit the site. Inspections of these areas must occur once every 7 days as well as within 24 hours of any storm event of greater than 0.5 inch. Inspection frequencies may be reduced to once every month for sites that have been finally or temporarily stablized, or during defined seasonal arid periods in arid or semiarid areas. The specific requirements for inspections specified in the general permit must be followed, and an appropriate report prepared. Reports must be maintained for at least three years from the date of final stabilization.

The SPPP must also specify the contractors and subcontractors that will implement each specific measure in the plan. All contractors and subcontractors must also, under penalty of law, certify that they understand the terms and conditions of the stormwater general permit. Records of all reports required under the permit, including SPPPs and all data used to compile the NOI, must be maintained for at least three years from the date of final stabilization. The permit does not contain any monitoring requirements or require the submission of any reports. As in the case of the industrial general permit, state section 401 certifications added some additional or different conditions to EPA's general permit. A permit is terminated by filing a notice of termination after the site has undergone final stabilization or the permittee has transferred operational control to another permittee. Final stabilization occurs when all soil-disturbing activities are completed and a uniform perennial vegetation cover with a density of 70% for unpaved areas and areas not covered by permanent structures has been established or equivalent stabilization measures have been employed.

THE MUNICIPAL STORMWATER PROGRAM

Under the WQA of 1987, EPA must issue NPDES permits to large (serving a population of 250,000 or more) and medium (serving a population of 100,000 or more but less than 250,000) municipal separate storm sewer systems (MS4s). Under the statute, the permits

- may be issued on a system- or jurisdiction-wide basis
- shall include a requirement to effectively prohibit non-stormwater discharges into the storm sewers
- must apply a new standard to "require controls to reduce the discharge of pollutants to the maximum extent practicable, including management practices, control techniques and design and engineering methods" (33 U.S.C.).

Large and medium MS4s are defined by EPA as those

- located in an incorporated place with a population of (i) 100,000 or more but less than 250,000 (medium MS4s) or (ii) 250,000 or more (large MS4s)
- located in counties with a population of 100,000 or more but less than 250,000 (medium MS4s) or located in counties with a population of 250,000 or more (large MS4s), excepting, in both instances, systems located in the incorporated places, townships, or towns within such counties or
- designated by EPA as part of a large or medium

MS4 due to the interrelationship between the discharges of the designated MS4 and a large or medium MS4 (40 CFR).

EPA has listed the jurisdictions that are covered under the definitions of medium and large MS4s in the permit application rule, and in Appendixes F, G, H, and I to 40 CFR Part 122. The listing was based on the 1980 census. As data from the 1990 census become available, EPA will revise the list.

To date, EPA has issued a rule with permit application requirements for MS4s only (55 *FR*). Permit applications for MS4s have two parts. Part 1 was due November 19, 1991 for large MS4s and May 18, 1992 for medium MS4s (40 CFR). Part 2 was due November 16, 1992 for large MS4s and May 17, 1993 for medium MS4s. EPA is supposed to issue final permits to large MS4s by October 1, 1993 and to medium MS4s by May 17, 1994.

The regulations define "municipal separate stormwater" as a conveyance or system of conveyances (including roads with drainage systems, municipal streets, catch basins, curbs, gutters, ditches, manmade channels, or storm drains) that is owned or operated by a state or local government entity and is designed or used for collecting or conveying stormwater and is not a combined sewer or part of POTW. Operators of combined sewer systems (systems designed as both sanitary sewers and storm sewers) or POTWs are excluded from the permit application requirement because these systems should already be permitted under existing NPDES programs.

Operators of medium and large MS4s may submit a jurisdiction- or system-wide permit application. Where more than one public entity owns or operates a municipal separate storm sewer system within a geographic area (including adjacent or interconnected MS4s), such operators may be co-applicants to the same application.

Part 1 of the MS4 permit application must include the following elements:

- *General Information*: information about the permit applicant or coapplicant
- *Legal Authority*: a description of the legal authority to control discharges to the MS4. If existing legal authority is inadequate to control pollutants in stormwater discharges, a plan to augment existing legal authority must be submitted
- *Source Identification*: detailed identification of all discharges to the MS4, including a topographic map, location of municipal outfalls, projected growth, location of structural controls, and location of waste disposal facilities
- *Discharge Characterization*: information characterizing the nature of system discharges, including existing quantitative data, the results of field screening analysis to detect illicit discharges, identification of receiving waters, a plan to characterize discharges from the system, and a plan to obtain representative data
- *Management Programs*: a description of the existing management programs to control the discharge of pollutants from the system, including a description of existing structural and nonstructural controls
- *Fiscal Resources*: a description of the financial resources currently available to complete Part 2 of the application.

Part 2 of the permit application is to provide information to supplement Part 1 and to provide municipalities with an opportunity to propose a program of structural and nonstructural measures to control the discharge of pollutants "to the maximum extent practicable." 55 *FR* 48045. Part 2 requires:

- a demonstration that the legal authority of the permit applicant to control discharges of pollutants to the system satisfies the criteria set out in 40 CFR § 122.26(d)(2)(i)
- supplementation of the source identification information submitted under Part 1, including identification of all "major outfalls" (a defined term) and an industrial activity description of each facility that may discharge stormwater associated with industrial activity to the system
- discharge characterization data from between 5 and 10 outfalls or field screening points designated by EPA or the state as representative of commercial, residential, and industrial land use activities based on Part 1 information. (The data shall also include estimates of the annual pollutant load of the cumulative discharges to waters of the United States; a proposed schedule to provide seasonal pollutant load and event mean concentration data for any constituent detected in any required sampling; and a proposed monitoring program for representative data collection for the term of the permit.)
- a proposed stormwater management program to reduce the discharge of pollutants to the maximum extent practicable
- assessment of controls, with estimated reductions in loadings of pollutants from system discharges resulting from the municipal stormwater management program
- fiscal analysis of capital, operation, and maintenance expenditures, by fiscal year for each year of the permit term, to complete the characterization data requirements, and stormwater management program requirements.

The cornerpiece of the Part 2 application is the proposed stormwater management plan (SWMP). This plan is to reduce the discharge of pollutants to the "maximum extent practicable using management practices, control techniques, and system design and engineering methods" (40 CFR). Under the regulations, proposed programs will be considered by the permit writer when developing permit conditions. The management plan allows the permittee to propose, in the first instance, the components of a program it believes are appropriate for preventing or controlling the discharge of pollutants (55 *FR*).

Stormwater management programs are to be based on

- a description of structural and source control measures to reduce pollutants from runoff from commercial and residential areas
- a description of a program, including a schedule, to detect and remove or permit illicit discharges and improper disposal in the system
- a program to monitor and control pollutants in stormwater from municipal landfills, hazardous waste treatment, disposal and recovery facilities, industrial facilities subject to section 313 of SARA Title III, and other municipal facilities identified as contributing a substantial pollutant loading to the system
- a program to implement and maintain structural and nonstructural best management practices to reduce pollutants from construction sites.

The stormwater management program must address illicit discharges to the system. The regulations define an illicit discharge as any discharge to an MS4 that is not composed entirely of stormwater, except discharges pursuant to an NPDES permit and discharges resulting from firefighting activities (40 CFR). However, as noted in the requirements for a SWMP, not all such flows must be prohibited. Certain types of flows such as landscape irrigation, groundwater infiltration, air conditioning condensate, lawn watering, and street wash waters need to be prohibited only when such flows are identified as significant sources of pollutants (55 *FR*).

PHASE II STORMWATER REGULATIONS

The 1987 Water Quality Act required EPA to establish a two-phased approach for the control of stormwater discharges. Phase I consists primarily of permitting stormwater discharges associated with industrial activity or stormwater discharges from "large" or "medium" municipal separate storm sewers. Phase II of the stormwater program covers all stormwater discharges not addressed under Phase I, and could include all other municipalities, as well as light industrial and commercial activities.

CWA Section 402(p)(5) requires EPA to conduct two studies on Phase II stormwater discharges. The first study will identify those classes of discharges that may be addressed in Phase II, and evaluate the nature and extent of pollutants in such discharges. The second study will evaluate procedures and methods to control Phase II stormwater discharges to the extent necessary to mitigate impacts on water quality. EPA has requested public comment on ways to implement the second phase of the stormwater permitting program for sources and activities not regulated under Phase I (57 *FR*). EPA specifically requested comment on a variety of options that range from comprehensive permitting of all municipal, light, industrial, and commercial activities that generate stormwater runoff to little or no NPDES permitting of Phase II sources. EPA's final approach to the Phase II stormwater permit program will be developed in the near future.

REFERENCES:

Moffa, Peter E., 1989. *Control and Treatment of Combined Sewer Overflows*, New York: Van Nostrand Reinhold.

The Federal Water Pollution Control Act, as amended, and commonly known as the Clean Water Act (CWA). 33 U.S.C. § 1251 *et seq.*

USEPA Combined Sewer Overflow (CSO) Control Policy (Draft, Sept., 1993).

U.S. EPA 1993, NPDES Storm Water Program Question and Answer Document: Vol. 2, U.S. EPA, Office of Water, Washington, DC, EPA 833-F-93-0028 at p. 4.

USEPA, Federal Register, National Pollutant Discharge Elimination System Permit System Application Regulations for Stormwater Discharges; Final Rule, Vol. 55, No. 222, November 16, 1990.

USEPA, Federal Register, National Pollutant Discharge Elimination System Applications Deadlines, General Permit Requirements and Reporting Requirements for Stormwater Discharges Associated with Industrial Activity; Final Rule, Vol. 57, No. 64, April 2, 1992.

USEPA, Federal Register, National Pollutant Discharge Elimination System; Stormwater Discharges; Permit Issuance and Permit Compliance Deadlines for Phase I Discharges; Final Rule, Vol. 57, No. 244, December 18, 1992.

USEPA, Federal Register, Water Pollution Control, NPDES General Permits and Fact Sheets: Storm Water Discharges From Industrial Activity; Notice, Vol. 58, No. 222, November 19, 1993.

Weiss, Kevin, and James Gallup, 1988. "Federal Requirements for Stormwater Management Programs." In: Proceedings *Design of Urban Runoff Quality Controls*, ed. Larry A. Roesner, Ben Urbonas, and Michael B. Sonnen, pp. 100–106, New York: ASCE.

2

Identifying the Problem: Simulating Stormwater Problems with Mathematical Models

OVERVIEW

This chapter introduces and discusses the use of deterministic mathematical models to simulate the movement of stormwater runoff and pollutants in watersheds, especially those in municipal and industrial settings. A typical municipal or industrial stormwater problem setting is shown in Figure 2.1. A well-conceived and validated model can be an extremely valuable tool for predicting such a complicated system's behavior under a variety of conditions. A model can answer questions that would be impractical to answer through field measurements and observations.

Several models are capable of adequately simulating stormwater. None are necessarily better than others; some just fit certain needs better than others. Some require little

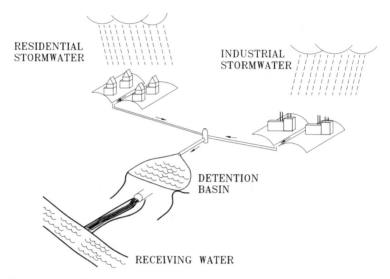

Figure 2.1.   Stormwater system.

more than a few readily available numbers, such as annual
rainfall and population density. At the other end of the
spectrum are models that require extensive data concerning the hydrologic characteristics of the watershed, the
hydraulics of the drainage system, and the pollutant
sources. The former is more appropriate for quick, preliminary estimates of stormwater volumes and pollutant
loads (usually on an annual basis). The latter is better
suited to detailed planning and design work. Each has its
place and, in fact, any study of a stormwater pollution
problem may make use of more than one model.

Why model and simulate industrial and municipal
stormwater discharges? The reason is simple. A model
that can accurately simulate stormwater discharges reduces the need to monitor the system physically and
allows us to simulate changes in the stormwater system
that would not be possible otherwise. Monitoring
stormwater is expensive and time-consuming. A model

allows us to use a limited number of samples (although often we try to get by with too few) to build a credible model that can essentially "fill in" for the times and places we didn't monitor *and* allow us more easily to make predictions for the future.

The use of any computer model to simulate a stormwater is inherently limited in a number of ways. First, a computer model cannot improve a data base. It can extract information from a data base; it cannot overcome data inadequacies. Second, no model will produce completely accurate results because every model is incomplete, biased, or just plain wrong in its representation of the system. Third, numbers produced by a computer model are no more accurate than numbers produced by hand calculations; they are simply produced faster. Placing a model on a computer will not improve it; you will just get the results, good or bad, faster.

All models have limitations and their use can be frustrating, especially for those that fail to recognize these limits. The wisest thing a model user can do is learn not to expect too much from a computer model. A model just provides a quicker way of executing a large number of calculations, all of which the user should understand and support as if he or she were doing the same calculations by hand.

Models are important to the assessment and abatement of stormwater pollution problems, but they're just tools and not surrogates for sound judgment and analysis. A model must be placed in its proper role in the overall analysis and design process and its output must also be interpreted with a keen awareness of the model's limitations and assumptions.

Much of what appears here was drawn from a companion book dealing with combined sewer overflows (Moffa et al., 1990) and another that covers stormwater modeling in far more detail (Nix, 1994). The reader is strongly encouraged to refer to the latter for a more rigorous presentation.

Definitions

Definitions vary in the field of stormwater modeling. Several are given below so that the remainder of the chapter will be clear.

System—A network of interacting components capable of responding to one or more stimuli (e.g., a watershed)

Model—Any representation of a system by something other than the system itself

Simulation—The application of a model (The terms "modeling" and "simulation" are often interchanged, but keeping them separate is sometimes useful. The construction of a model is "modeling." Operation of the model to mimic the behavior of a system is "simulation.")

Mathematical model—A model using mathematical relationships and algorithms to represent a system

Computer model—A mathematical model in which the mathematical operations are carried out on a computer

This chapter will deal with mathematical models, and principally computer models. There are generally two different kinds of mathematical models:

Deterministic model—A mathematical model built on

an understanding of the system and the interaction between system inputs (e.g., precipitation) and system outputs (e.g., runoff)

Stochastic model—A mathematical model designed to produce a series of random responses that have the statistical characteristics of the historical responses of the system (in other words, a model that does not focus on understanding the interaction of inputs and outputs through a system, but on the characteristics of the output).

This chapter will deal with deterministic models. There are generally two different kinds of deterministic models:

Distributed model—A deterministic model in which spatial variabilities in watershed characteristics are continuous functions of the position in the watershed

Lumped deterministic model—A deterministic model in which each watershed characteristic is constant over the watershed.

Nearly all stormwater models are lumped models that can be individually applied to portions of the watershed and the drainage system to create the impression of a distributed model. That is, each portion will have uniform spatial characteristics but the various portions pieced together create spatial variability.

Other model definitions include:

Run—A single execution of a model

Run time—The time needed to execute a model run on a computer

Simulation time—The time simulated by a model run

Variable—A quantity that can vary in a particular model run

Parameter—A quantity that cannot vary in a particular model run (However, the reader should note that the definitions for *variables* and *parameters* are not as distinct as one might think. Time essentially divides the two. It is quite conceivable that a parameter—e.g., percent impervious area—will become a variable, given enough time.)

Algorithm—A calculation procedure

Single-event simulation—The use of a model to simulate the response of a catchment to a single storm event (normally a "design" event)

Continuous simulation—The use of a model to simulate the response of a catchment to a series of storm events *and* the hydrologic processes that occur between them.

Hydrologic and water quality definitions include:

Catchment or watershed—The area producing the runoff passing a particular channel or stream location (e.g., gaging station)

Subcatchment or subwatershed—A portion of a catchment or watershed producing the runoff passing a channel or stream location upstream of the location defining the catchment or watershed

Precipitation depth—The volume of precipitation occurring over a specified area expressed as an equivalent depth over that area (e.g., inches or millimeters)

Rainfall intensity—The rate at which rainfall occurs

over an area, expressed as an equivalent depth per unit time over that area (e.g., inches/hour or millimeters/hour)

Hyetograph—A plot of rainfall intensity versus time (see Figure 2.2)

Runoff volume—The quantity of runoff from an area, expressed as a volume (e.g., cubic feet, acre-feet, or cubic meters)

Runoff depth—The volume of runoff discharged from an area, expressed in terms of an equivalent depth over that area (e.g., inches or millimeters)

Runoff rate—The volume or depth per unit time (cubic feet/second, cubic meters/second, or acre-feet/hour for volume/time; inches/hour or millimeters/hour for depth/time) at which runoff occurs from an area

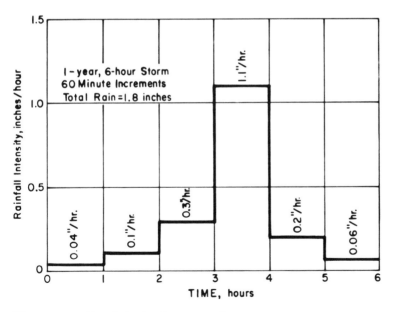

Figure 2.2. Precipitation hyetograph.

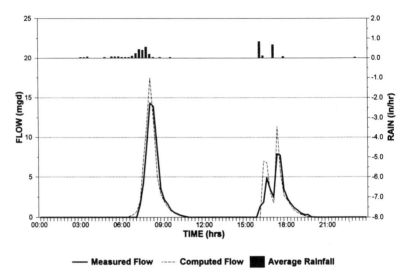

Figure 2.3.   Runoff hydrograph.

Hydrograph—A plot of the runoff rate versus time (see Figure 2.3)

Pollutant load—The quantity of a pollutant discharged from an area, expressed as a mass or mass per unit area (e.g., pounds, kilograms; pounds/acre, kilograms/hectare)

Pollutant load rate—The pollutant load discharged from an area per unit time (e.g., pounds/day, kilograms/day; pounds/acre-day, kilograms/hectare-day)

Pollutant concentration—The ratio of pollutant mass (or occasionally another measure of quantity, such as number of coliform bacteria) to volume (e.g., pounds per cubic foot or milligrams per liter)

Pollutograph—A plot of pollutant concentration (or sometimes the pollutant load rate) versus time (see Figure 2.4)

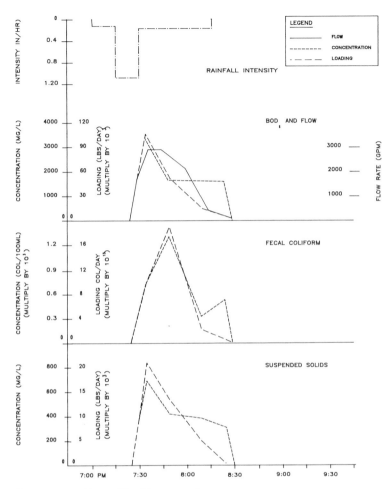

Figure 2.4. Runoff pollutograph.

SELECTION OF A MODEL

The assumption here is that the reader is not interested in developing a model, but rather in selecting a good "off-the-shelf" model. This selection decision should be based on what the model is expected to accomplished. A model designed to provide rough estimates will be of little use in design work. Conversely, a model intended for design work will probably be too involved for simple preliminary study.

In theory, the selection of a model can be based on a careful benefit–cost analysis. As in all engineering work, the identification and quantification of benefits and costs is difficult. But the fundamental concept will help immensely, even if the analysis is more qualitative than quantitative.

Figure 2.5 presents the basic concept of benefit–cost analysis. In general, the more complex a model, the more information it is capable of producing. (This may or may not be true; but certainly, no model is intentionally given additional detail and complexity to produce less information.) However, the cost of that information is rising rapidly. What are the benefits of that information? More information is good, but as we obtain more, the next piece of information provides less additional value. What this implies, of course, is that there is a model which will maximize the net gain or benefits (see Figure 2.5).

The concepts of benefit–cost analysis can be used to make a good point: a model should be selected with much the same care as one would apply to any engineering decision. However, the decision is more complicated than the selection of a model to meet the technical goals of a particular application. The following are other important factors to consider.

1. *Availability of suitable hardware.* The increasing power and decreasing costs of personal computers have given scientists and engineers ready access to the computing resources needed to operate a wide array of models. Nearly all models are available in forms suitable for personal computers. However, it is still true that some models are not suitable for certain computers.

2. *Availability of trained personnel.* The model user must

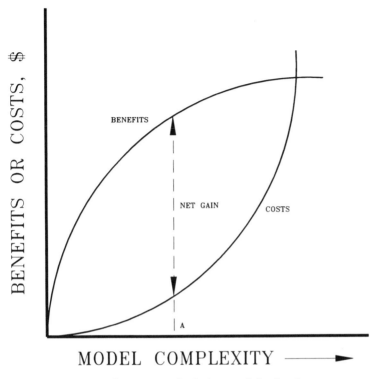

BENEFITS OR COSTS, $

BENEFITS

NET GAIN

COSTS

A

MODEL COMPLEXITY ⟶

Figure 2.5. Benefit–cost analysis for model selection.

be competent enough to know when the model is not producing reasonable results. Some view models (and most computer software) as crutches: there to fill in for the weaknesses of the user. That view is dangerous and unethical. The user is always responsible for the model and the interpretation of the results.

3. *Long-term commitment to the model.* At times it may be desirable to develop expertise with a particular model even if that model is not ideal for the current application. If a number of similar applications are anticipated it may be more beneficial to invest heavily in one model than to switch models from project to project. However, this should be done without grossly violating the first rule of model selection: fit the model to the task at hand. Many models can be tailored to serve a number of roles.

4. *In-house model experience.* Experience with a particular model may already exist in a project team. In this case, the fact that little or no training or "warm-up" is necessary may overwhelm some of the minor deficiencies of that model.

5. *Acceptance and support of the model.* It is quite possible that a fairly obscure model will seem to meet the needs of a given project perfectly. However, there is no substitute for learning from the experiences of colleagues. If the model is used by few, it becomes that much more difficult to establish its credibility and to interpret its results. Likewise, if there is little or no support from the model developers (e.g., to provide updates, correct errors, provide user assistance) the chances for success are diminished. Large organizations with a substantial commitment to modeling may be able to provide such support internally, but many smaller organizations would find this difficult. A well-established model with a large user community and extensive support that meets project needs somewhat less than ideally may be the better choice.

There are, to be sure, other factors to be considered. It should be clear, though, that the selection of a model is not a trivial exercise.

Catalog of Models

There has been plenty of activity in stormwater modeling over the last three decades and there are plenty of models from which to choose. These models can be classified in a several ways, but it is probably most useful to classify them by the level of *routing* supported by the model.

Routing refers to the movement of runoff and pollutants from one point in the watershed to another. The level to which this is done varies considerably among models. Three levels are identified in Table 2.1 and discussed in more detail below.

At the "low" end of the scale are very simple models that do not route runoff and pollutants through the watershed at all. Typically simulated runoff and pollutant loads are generated *at the outlet* of the watershed through a simple relationship to rainfall. Because there is no attempt to account for the delaying effect of the watershed and its drainage system, the time distribution of runoff is not particularly accurate. However, the runoff volume can be reasonably estimated. These models usually require a minimal amount of input data, and often need little, if any, help from a computer. Their principal purpose is to provide a "first-cut" assessment of the problem and perhaps a crude abatement strategy. Examples include SWMM Level I (Heaney et al., 1976; Heaney and Nix, 1977) and statistically based procedures developed by Howard (1976), DiToro and Small (1979), and Hydroscience, Inc. (1979).

At the "middle" level of sophistication are models that run on computers and use relatively simple methods to route flow and pollutants through sewers, channels, and/or storage basins. Simple routing models attempt to generate simulated stormwater at a point other than the watershed outlet and to route those flows through a simple drainage or storage network to the watershed outlet. The result is a more accurate picture of the time distribution of runoff. These models, because of their relative simplicity, are often designed to simulate stormwater runoff over long periods (e.g., continuous simulation). Excellent examples of this type include the Corps of Engineers'

Table 2.1. Characteristics of Represnetative Urban Runoff Models (Nix, 1991)

Model	Routing Level	Time Domain	Predictive Method	Pollutant Transport
SWMM-Level I	none	long-term averages	empirical	none
Statistical	none	long-term averages	frequency distribution	none
STORM	simple storage	continuous	buildup and washoff, soil loss	mass balance
HSPF	simple channel	continuous and single event	buildup and washoff, soil loss	completely mixed
QQS	channel and pipe routing	continuous and single event	buildup and washoff	plug flow
TR55	simple storage	single event	frequency distribution	none
SWWM	channel and pipe routing	continuous and single event	buildup and washoff, soil loss and rating curve	completely mixed, plug flow (in storage)

Storage, Treatment, Overflow, Runoff Model, or STORM (Hydrologic Engineering Center, 1977), and the Hydrological Simulation Program–Fortran, or HSPF (Johanson et al., 1984).

At the "high" end are models capable of routing flows through a fairly complete depiction of the watershed's drainage system and producing hydrographs and pollutographs at numerous locations. These models contain relatively sophisticated hydraulic routing algorithms suitable for design work. Some are better suited for single-event simulation (e.g., because the runoff generator may not be designed to account for interevent processes) and others are capable of continuous simulation (although lengthy run times and voluminous output often make it unattractive). Some of the better-known models of this type include the EPA Storm Water Management Model, or SWMM (Huber and Dickinson, 1988), the Illinois Urban Drainage Simulator, or ILLUDAS (Terstriep and Stall, 1974), and the Quantity–Quality Simulator, or QQS (Geiger and Dorsch, 1980).

This classification scheme is useful because the detail to which a model routes flow and pollutants through a watershed says a great deal about the kind of work for which the model is best suited. If we are interested only in runoff volumes and loads at the watershed outlet then a model that routes flows through sewers and drainage channels is probably unnecessary. However, if we are interested in how runoff and pollutants occur over time, the storage effect provided by the watershed should be analyzed with a more sophisticated model.

A number of compendiums and comparisons of models have appeared over the last several years (American Public Works Association, 1981; Brandstetter, 1977; Chu and Bowers, 1977; Hall, 1984; Huber, 1985, 1986; Huber and Heaney, 1980, 1982; Nix, 1991; Nix, 1994; U.S. Environmental Protection Agency, 1983; Whipple et al., 1983). The summaries by Huber and Heaney (1980, 1982) are particularly useful. The reader should refer to these or one of the other works for a more detailed discussion of the models available for stormwater modeling.

As discussed earlier, the type of model selected has much to do with the modeling objectives. Preliminary studies and assessments can probably be conducted with simple, nonrouting models. Planning studies will most likely find the simpler routing models of the most value. Design work can make use of the more complete routing models. But it's useful to note that some of the more sophisticated models are capable of performing simpler types of analyses. In addition, some of the simpler models can guide the use of more-sophisticated models by providing some preliminary results. So it is quite possible that a sequence of models may be most appropriate or that one model can be operated at various degrees of detail and complexity to serve many needs.

BUILDING A CREDIBLE MODEL

All stormwater models are imperfect representations of the physical system. That is true because our models can be no better than our incomplete understanding of those systems. Therefore, it becomes necessary to test the algorithms for proper functioning and to adjust parameters so that the model adequately represents the physical system. The former process is known as verification; the latter is known as validation. Many also refer to the process of validation as calibration or parameter optimization.

A third technique is useful for both model credibility and the analysis of system behavior. Sensitivity analysis is used to determine which model parameters are important—that is, which have a significant impact on model output. A fourth technique is the constant review of the output from the production runs (the runs of the validated model used to meet the objectives of the stormwater modeling study).

Verification

The task of verification is sometimes overlooked. That is unfortunate because much can be learned about model behavior, with a minimal investment of effort.

The verification of a model involves running a "check-out" of its components to see whether they perform as expected and to learn the model's idiosyncrasies. An example of a simple verification exercise is illustrated in Figure 2.6. This figure shows the well-known response of a catchment to steady rainfall. Obviously, a model that does not respond in a similar manner to a constant rainfall is not properly constructed. This is an overly simple example, but other checks on the response of the model to controlled and well-understood stimuli can shed light on problems and build confidence in the model. These checks can also help the user understand the nuances of the model and, as a result, make the later validation effort (as well as the "production" runs) more efficient.

Validation

Validation refers to the process of collecting data to describe the inputs and outputs to a watershed for a wide range of conditions *and* adjusting the model parameters so that the model adequately replicates the watershed as characterized by these data. The art of model validation is indeed just that—an art. In spite of attempts to formalize and automate validation procedures over recent years, the process is still very much an iterative, heuristic exercise

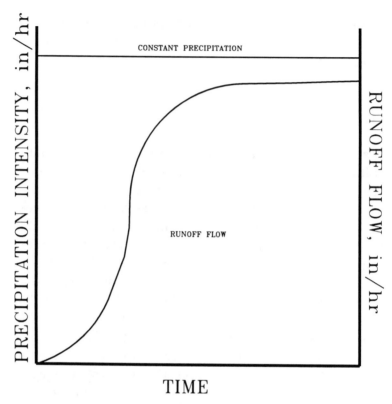

Figure 2.6.   Verification exercise—response of a catchment to
steady rainfall.

involving repeated model runs and subsequent adjustments. However, if the user has made the small investment in time to understand the model in the verification
stage, he or she will be able to make adjustments in an
intelligent, efficient manner.

One of the most perplexing problems facing the model
user is the development of the data base against which the
model is going to be validated. How many storm events
should be monitored? How many rain gages are needed
and where should they be placed? Where in the watershed
should flow and quality be monitored? The answers to

these questions can be provided only by a careful review of the modeling objectives and the available resources (including time, money, personnel, and previously collected data). However, it is crucial that neither the modeling work nor the monitoring program operate in a vacuum. The monitoring and modeling results should complement *and* guide each other. As the study progresses, both efforts will produce valuable insights that may affect the other.

Any validation exercise should explore as wide a range of conditions as practical. In other words, calibrate the model for several storms and for several locations in the system. It is likely that a model will respond to different conditions in a somewhat inconsistent manner and the key is to calibrate the model over the range of *pertinent* conditions. For example, a model "calibrated" for a small storm may perform poorly for a large storm and vice versa. But if the modeling effort is intended to provide information for design work (where large storms would be of interest) then it is probably not so important for the model to perform well for very small storms.

With the previously discussed points in mind, there are some general steps that the user of any model should consider when calibrating a model to a stormwater drainage system and its catchment.

1. *Identify the important model algorithms and parameters.* For a given application, a particular model presents the user with several pertinent and not-so-pertinent algorithms and parameters. For example, parameters associated with an infiltration routine are not very important for the user to identify these algorithms and parameters so that resources are not wasted on them (e.g., to estimate infiltration parameters in the highly urban watershed). This will also simplify the validation

process. Sensitivity analysis is very useful in this effort (see later discussion).

2.   *Classify the parameters.* Model parameters range from those that are almost entirely conceptual and, as a result, very difficult to estimate (e.g., width of overland flow) to those that are well defined and relatively easy to measure (e.g., area). For the most part, it is unwise to spend a great deal of effort to establish *a priori* estimates for the more conceptual parameters. Essentially, these parameters are the prime validation or calibration "knobs." These are the parameters that should be looked to first when calibrating the model. The parameters that are more directly tied to measurable quantities should be carefully estimated before the validation process and then left alone. The more conceptual parameters should be estimated in a considered and consistent manner before model validation, but one should not waste too much time trying to "measure" them. Some help can be obtained by a review of the literature to identify applications of the model to other watersheds.

3.   *Adjust the "volume" parameters.* It is probably better to adjust the parameters most responsible for producing the runoff volume first. Without a fairly accurate representation of the volume of runoff produced by a storm or storms the rest is fairly meaningless. How does one know which are most responsible for volume? Again, one knows this by understanding how the model represents the physical world and by sensitivity analysis.

4.   *Adjust the "peak and shape" parameters.* The peak and shape of the runoff hydrographs should be adjusted after the simulated volumes agree reasonably well with the measured data. The parameters responsible for hydrograph peaks and shape (e.g., recession) are, again, discovered through an understanding of the model and sensitivity analysis. Bear in mind, however, that the

effect of changing the value of one of these parameters may alter the volume results to some degree, and other parameters may have to be readjusted in what is actually an iterative process.

5. *Adjust the water quality parameters.* A model that is not first calibrated for stormwater quantity cannot produce reliable stormwater quality results. Therefore, parameters responsible for generating pollutant loads and their transport through the watershed should be adjusted after the values for the quantity parameters are felt to be adequate.

What does it mean for a model to be "calibrated"? This is difficult to answer. The process of calibration is used to build confidence in the model to act as a predictive tool. A model is calibrated when it is capable of predicting at the accuracy needed. There are no arbitrary standards, nor is there a need for any. The user is ultimately responsible for his or her work and, thus, must be satisfied that the model will provide the information needed.

Although there are more-sophisticated methods, most calibration efforts will be visual (e.g., by comparisons of actual and simulated hydrographs) and somewhat subjective. Typically, the user will visually inspect the accuracy of the results from one run, make educated adjustments, execute another run, and end the process when satisfactory results are obtained.

Sensitivity Analysis

Sensitivity analysis is used to examine the sensitivity of the model to changes in parameters. This analysis has at least two very beneficial uses. First, as discussed above, it is important to know which parameters have the greatest effect on various aspects of the model output so that the user knows which ones to adjust in the calibration process.

Second, after the model has been calibrated and is being used to make predictive estimates, it is very useful to be able to examine the effects of possible changes in parameter values, *especially* those that were used in the calibration process. Why? Because those were the parameters about which the user is probably the most unsure. Varying a parameter within a range of possible error and documenting the effect on the output will help to quantify the effect of the uncertainty associated with that parameter.

Figures 2.7 and 2.8 show a very effective way to study the sensitivity of a model to different parameters in terms of the model output. In these examples, changes in several parameters are depicted by a percent change (on the abscissa) from the calibrated values and the effects are shown as percent changes in an output from those obtained by the calibrated model. Figure 2.7 shows the technique applied to runoff volume and Figure 2.8 demonstrates the same for the peak runoff rate. These figures effectively show the parameters that have the most impact on model output.

Monitoring the Production Runs

The credibility of a model can be enhanced further by continually monitoring the output of the production runs. These runs may produce suspicious results that call for a reevaluation of the model parameters. In truth, the calibration process is never finished.

OTHER MODELING ISSUES

Aside from the issues raised in the previous discussions, there are a number of other practical questions that every

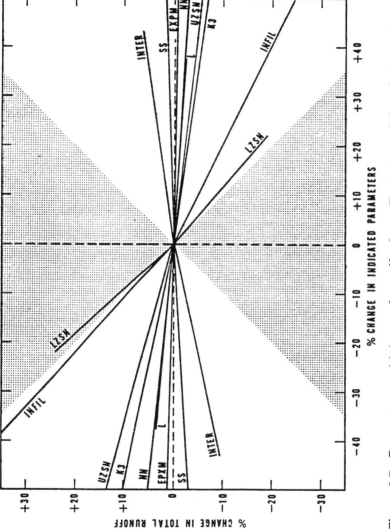

Figure 2.7. Parameter sensitivity, total runoff volume (Donigian and Crawford, 1976).

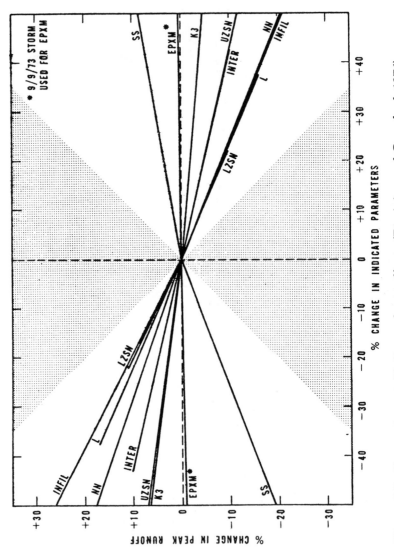

Figure 2.8.   Parameter sensitivity, peak runoff rate (Donigian and Crawford, 1976).

model user should consider. These involve the basic conditions under which the model will be defined and executed.

Single-Event, Multiple-Event, and Continuous Simulation

The question of whether a modeling study should focus on a single "design" event, multiple, but discrete, "design" events, or a continuous series of events cannot be answered here. This is a question that must be addressed in a larger context. Nevertheless, there are some observations that may be useful.

1. Parameters adjusted through a calibration exercise for a single-event simulation may not be applicable to a continuous simulation for the same catchment. The effect of certain parameters on model output can be different from storm to storm.

2. The use of a "design" rainstorm (e.g., a 10-year storm) is fundamentally flawed because the resulting runoff event does not necessarily have the same return period. This is principally due to the fact that antecedent conditions have a great deal of influence on the severity of a particular storm. For example, a 2-year rainstorm may result in a 5- or 10-year runoff event if the ground is already near saturation. Continuous simulation can produce the response of a catchment to many, many years of rainfall conditions. From these responses the user can identify critical runoff events that may be suitable for "design" storm simulations.

3. Continuous simulation lessens the importance of estimating the condition catchment at the start of the simulation. In single-event simulation the initial condition of the catchment (soil moisture, water in depression storage, etc.) can have considerable effect on the model output (and, of course, the actual catchment). A long, continuous simulation (say a year) would be affected only for the first few events and the overall impact would probably be minimal.

The Time Step

Many stormwater models rely on numerical approximations to solve one or more differential and/or integral equations. Take, for example, the simple continuity-of-mass equation, i.e.,

$$\frac{dM}{dt} = m_{in} - m_{out} \qquad (2\text{-}1)$$

where
$\quad M$ = mass stored in the system
$\quad t$ = time
$\quad m_{in}$ = mass inflow rate, mass/time
$\quad m_{out}$ = mass outflow rate, mass/time

A numerical approximation of this differential equation would involve replacing dM/dt with $\Delta M/\Delta t$. The term t is what is known as the "time step" or "integration step." Some models fix the time step; others allow the user to select it. In the latter case, the user should follow the guidelines provided by the model developers, but the effect of different time steps on model output should be investigated, much as for any other model parameter.

In general, the time-step value affects two things: (1) the ability of the model to respond to rapidly changing conditions and (2) the stability of the numerical solution procedures. The former is demonstrated in Figure 2.9. The hydrograph labeled *A* is rapidly changing; in contrast, the hydrograph labeled *B* is much less dynamic. Hydrograph *A* will never be reproduced by a model using the long time step Δ_t' whereas the relatively short time step Δ_t'' will allow the numerical procedures to respond to the rapidly fluctuating conditions. However, the longer time step may well be adequate for hydrograph *B*. In more general terms, the time step should be consistent with the watershed's response time.

The stability of a model is sometimes adversely affected by a time step that is too large. A large time step will often cause a numerical procedure to grossly "overshoot" or "undershoot" the true solution because of the reliance of many numerical procedures on recent information (e.g., from the last time step) to project the output value in the next time step. If the time step is too large, rapid changes

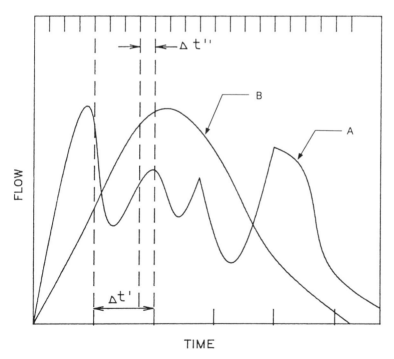

Figure 2.9. Selection of an appropriate time step.

in input conditions will not be properly reflected in the output and the numerical procedure will probably become unstable attempting to reconcile the conflicting information.

The time step is one of the principal determinants of a model's run time. A good time step will simultaneously maintain stability and keep the run time reasonable. However, computers are becoming more powerful at a dizzying rate and excessive run times are becoming a fairly small concern. Nevertheless, the determination of an appropriate time step still deserves some attention.

Spatial Detail

Nearly all stormwater computer models are, in truth, lumped models. However, changes in the physical characteristics of the catchment can be modeled by assuming that they are constant over smaller portions of the catchment—i.e., subcatchments. This is in contrast to models that include the physical heterogeneities (slope, for example) in the governing differential equations. These are distributed models. Lumped models are, of course, much easier to construct and they can be made to act more and more like distributed models if the catchment (and its drainage system) is divided into more and more subcatchments. Of course, more detail or *discretization* requires more effort.

In general, watersheds that are heterogeneous should be modeled with more detail than relatively homogeneous watersheds. But additional detail that adds nothing in terms of useful information is wasteful. The most desirable level of spatial discretization is strongly tied to the modeling objectives and how the model responds to different levels of discretization.

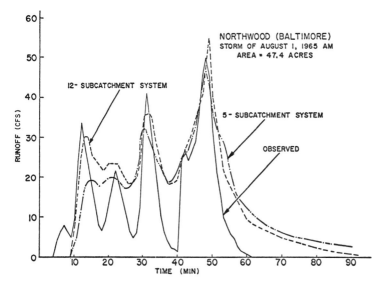

Figure 2.10. Effect of different discretization levels, Northwood catchment, Baltimore, Maryland (Metcalf and Eddy, Inc. et al., 1971).

An example is useful. Figure 2.10 demonstrates the response of the EPA Storm Water Management Model (see later discussion) to various discretization schemes for a catchment in Baltimore, Maryland. The runoff hydrographs produced by the 12-subcatchment and 5-subcatchment schemes are not very different. In this case, little seems to be gained by breaking the catchment into 12 subcatchments. Would five or less be adequate? Perhaps. The modeling objectives would, again, drive the decision. For example, if the objective is to study the effect of changes in a number of smaller sewers, then the user will want to include that level of detail. Alternatively, if the objective is simply to predict runoff hydrographs at the outlet of the watershed, then the inclusion of a lot of small sewers and small homogenous subwatersheds would be unnecessary.

The State of Stormwater Quality Simulation

Sonnen (1980) claimed that the state of the art in modeling stormwater quality was fairly dismal. It is not clear that much has changed. Most stormwater models can be successfully calibrated for runoff quantity. The same cannot be said for runoff quality, at least not so easily. Cases in which the actual and simulated water quality data differ by several times are not uncommon. The complex chemical, biological, and physical processes have successfully defied efforts to reduce them to mathematical statements!

The rather poor state of the art is belied by the large number of parameters and complicated algorithms found in the stormwater quality sections of some larger models. The level of complication implies rigor that isn't really there. Users should study stormwater quality algorithms very carefully (with the help of verification and sensitivity analysis) and apply them with good judgment.

Modeling studies and monitoring programs should be carefully coordinated, especially when dealing with stormwater quality. Stormwater runoff can be simulated fairly well by most models without extraordinary effort. But stormwater quality can be successfully simulated only with a solid data base. Simulation is complementary to monitoring, but it in this case the stormwater quality algorithms are not able to fill as much of the void as their stormwater quantity counterparts.

EPA STORM WATER MANAGEMENT MODEL

Overview

There are a number of models capable of simulating stormwater from municipalities and industrial sites. Each has strengths and weaknesses that determine its suitability for specific applications. However, the Environmental Protection Agency Storm Water Management Model (SWMM) is probably the most popular. What follows is a mostly qualitative description of how SWMM works. This serves two purposes. One is to give the reader an insightful overview to a very popular tool that is hard to get from the model documentation. The other is to serve as a good introduction to the more sophisticated stormwater models. SWMM has served as the benchmark for over two decades and a study of its inner workings will provide a good look into several others. SWMM is flawed, but overall it performs well.

SWMM is a large, complex model, capable of simulating the movement of precipitation and pollutants from the ground surface, through pipe and channel networks and storage/treatment units, and finally to receiving waters. The model may be run for a single event or on a continuous basis for extended periods of simulation time.

SWMM has been released under several different "official" versions (Metcalf and Eddy, Inc. et al., 1971; Huber et al., 1975, 1984; Huber and Dickinson, 1988; Roesner et al., 1988), and there are many "unofficial" versions modified for specific purposes. The official versions were primarily designed for mainframe use, but the later versions can be executed on a personal computer. The current version of SWMM (Version 4) is available on floppy disks and may be obtained (along with the documentation) from the Center for Exposure Assessment Modeling (CEAM), U.S. Environmental Protection Agency, College Station Road, Athens, Georgia 30613. The cost is minimal. There is an extensive body of literature describing its use (Huber et al., 1985) and a user's group dedicated to SWMM and stormwater modeling in general (contact CEAM in Athens, Georgia). This large body of experience is an advantage that SWMM enjoys over all other similar stormwater models.

General Model Structure

SWMM is divided into several "blocks," each designed to handle a separate phase of the stormwater runoff process. Some of the these blocks—i.e., RUNOFF, TRANSPORT, STORAGE/TREATMENT, and EXTRAN—are computational blocks responsible for the hydrologic, pollutant generation and transport, and hydraulic calculations. Others blocks perform various service functions. A general operational schematic of SWMM is shown in Figure 2.11 and an overview is presented in Table 2.2. The tabular sum-

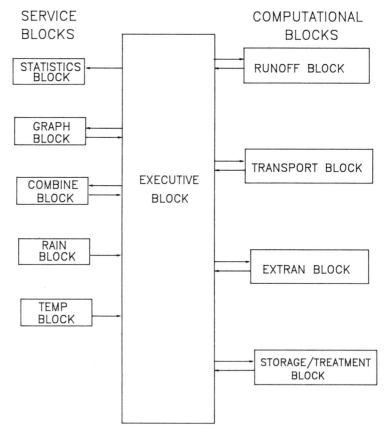

Figure 2.11. Operational schematic of the Storm Water Management Model (Huber et al., 1988).

Table 2.2. Summary of EPA Stormwater Management Model (SWMM) Characteristics (Adapted from Huber et al., 1988)

Applicable Land Drainage Area
(1) Urban (2) General nonurban
Time Properties
(1) Single-event or continuous simulation; both modes have an unlimited number of time steps. (2) Precipitation: input at arbitrary time intervals for single-event simulation (typically 1–15 min) and continuous simulation (typically 1-hr); for snowmelt daily max-min temperatures required for continuous simulation, temperatures at arbitrary intervals for single events. (3) Output at time-step intervals (or multiples); daily, monthly, annual, and total summaries for continuous simulation. (4) Time step arbitrary for single-event (typically 5 min) and continuous (typically 1 hr) simulations; variable time step available in RUNOFF Block; time step for Extended Transport Block (EXTRAN) routing depends on stability criteria; may be as small as a few seconds.
Space Properties
(1) Small to large multiple catchments. (2) Surface: lumped simulation of surface flow with allowance for up to 200 subcatchments and 10 input hyetographs, up to 200 channel/pipes may be simulated by nonlinear reservoir routing. (3) Channel/pipes: one-dimensional network, up to 200 conduit/nonconduit elements for TRANSPORT Block, up to 200 conduits in EXTRAN Block, up to 30 in-line storage units in TRANSPORT Block. Values easily changed using FORTRAN PARAMETER statement. (4) Catchment area may be disaggregated and modeled sequentially for simulation of areas too large for existing SWMM dimensions. (5) Storage/treatment simulated separately, receiving input from upstream routing. (6) Output from surface, channel/pipe, or storage/treatment simulation may serve as new input for further simulation by same or different blocks.
Physical Processes
(1) Flow derived from precipitation and/or snowmelt; snow accumulation and melt simulated using temperature-index methods developed by National Weather Service; snow redistribution (e.g., plowing, removal) may be simulated. (2) Overland flow by nonlinear reservoir using Manning's equation and lumped continuity, depression storage, integrated Horton or Green-Ampt infiltration (with optional subsurface routing),

recovery of depression storage via evaporation between storms during continuous simulation, also exponential recovery of infiltration capacity. (3) Subsurface routing only of flows through unsaturated and saturated zones simulated using lumped storages; subsurface outflow by power equation; simulation of evapotranspiration and water-table fluctuation. (4) Channel/pipes: (a) nonlinear reservoir formulation for channel/pipes in RUNOFF Block, includes translation and attenuation effects; (b) modified kinematic wave formulation in original TRANSPORT Block assumes cascade of conduits, cannot simulate backwater over more than one conduit length, surcharging handled by storing water at surcharged junction pending available flow capacity; (c) EXTRAN Block solves complete Saint-Venant equations including effects of backwater, flow reversal, surcharging, looped connection, pressure flow; (d) infiltration and dry-weather flow may enter conduit of either transport simulation. (5) Storage routing using modified Puls method assuming horizontal water surface; outlets include pumps, weirs, orifices. (6) Surface quality on basis of linear or nonlinear buildup of dust/dirt or other constituents during dry weather and associated pollutant fractions, power-exponential washoff with decay parameter a power function of low rate only (rating curve); erosion by Universal Soil Loss Equation. (7) Dry-weather flow quantity and quality on basis of diurnal and daily variation, population density, and other demographic parameters, buildup of suspended solids in conduits by dry-weather deposition using Shield's criterion. (8) Quality routing by advection and mixing in conduits and by plug flow or complete mixing in storage units, scour, and deposition of suspended solids in conduits (original TRANSPORT Block) using Shield's criterion. (9) Storage/treatment facility simulated as series-parallel network of units, each with optional storage routing. (10) Treatment simulation: (a) use of arbitrary user-supplied removal equations (e.g., removal as exponential function of residence time); (b) use of sedimentation theory coupled with particle-size–specific-gravity distribution for constituents.

Chemical Processes

(1) Ten arbitrary conservative constituents in RUNOFF Block, rainfall quality included, choice of concentration units is arbitrary; erosion of sediment is optional. (2) Four constituents may be routed through the original TRANSPORT Block (with optional first-order decay), three through the

STORAGE/TREATMENT Block, and none through EXTRAN (quantity only).

Biological Processes

(1) Coliform washoff may be included. (2) Biological treatment may be simulated.

Economic Analysis

Amortized capital plus operation and maintenance costs for control units are determined.

Mathematical Properties

(1) Physically based model. (2) Surface quantity: iterative solution of coupled continuity and Manning equations, Green-Ampt or integrated form of Horton infiltration (infiltration rate proportional to cumulative infiltration, not time). (3) Surface channel/pipe routing: nonlinear reservoir assuming water surface parallel to invert. (4) Channel/pipes: (a) TRANSPORT: implicit finite difference solution to modified kinematic wave equation; (b) EXTRAN: explicit finite difference solution to complete Saint-Venant equations, stability may require short time-step. (5) Storage/detention: modified Puls method requires table lookup for calculation of outflow. (6) Surface quality, quality routing, and treatment: algebraic equations, no iterations required once flows and conduit volumes are known.

Computational Status

(1) Coded in FORTRAN-77, approximately 25,000 statements long. (2) Has been run on IBM, UNIVAC, CDC, Amdahl, VAX, Prime, Harris, Boroughs, and other mainframe and minicomputers; mainframe version must be compiled. (3) Microcomputer version for IBM-PC compatibles uses same FORTRAN code; requires 512K bytes plus math coprocessor; hard disk desirable; executable and source code available. (4) May be run in modular form (RUNOFF, TRANSPORT, EXTRAN, and/or STORAGE/TREATMENT, plus executive and service routines, e.g., plotting, file combining, statistics, data input). (5) Largest block requires about 125,000 words or 500K bytes of storage. (6) Available on a magnetic tape or floppy disks. (7) Requires up to eight off-line storage files.

(1) Historical or synthetic precipitation record; uses National Weather Service precipitation tapes for continuous simulation. (2) Monthly or daily evaporation rates. (3) For snowmelt: daily max-min (continuous) or time-step (single-event) temperatures, monthly wind speeds, melt coefficients and base melt

temperatures, snow distribution fractions and areal depletion curves (continuous only), other melt parameters. **(4)** Surface quantity: area, imperviousness, slope, width, depression storage, and Manning's roughness for pervious and impervious areas; Horton or Green-Ampt infiltration parameters. **(5)** Subsurface quantity: porosity, field capacity, wilting point, hydraulic conductivity, initial water-table elevation, ET parameters; coefficients for groundwater outflow as function of stage- and tail-water elevation. **(6)** Channel/pipe quantity: linkages, shape, slope, length, Manning's roughness; EXTRAN also requires invert and ground elevation, storage volumes at manholes and other structures; geometric and hydraulic parameters for weirs, pumps, orifices, storages, etc.; infiltration rate into conduits. **(7)** Storage/sedimentation quantity: stage-area-volume-outflow relationship, hydraulic characteristics of outflows. **(8)** Surface quality (Note: Several parameters are optional, depending upon methods used): land use; total curb length; catch-basin volume and initial pollutant concentrations; street sweeping interval, efficiency, and availability factor; dry days prior to initial precipitation; dust/dirt and/or pollutant fraction parameters for each land-use or pollutant rating curve coefficient; initial pollutant surface loadings; exponential and power washoff coefficients; erosion parameters for Universal Soil Loss Equation, if simulated. **(9)** Dry-weather flow constant or on basis of diurnal and daily quantity/quality variations, population density, other demographic parameters. **(10)** Optional particle-size distribution, Shields parameter and decay coefficients for channel/pipe quality routing and scour/deposition routine. **(11)** Storage/treatment: parameters defining pollutant removal equations; parameters for individual treatment options, e.g., particle-size distribution, maximum flow rates, size of unit, outflow characteristics, optional dryweather flow data when using continuous simulation. **(12)** Storage/treatment costs: parameters for capital and operation and maintenance costs as function of flows, volumes, and operating time. **(13)** Data requirements for individual blocks much less than for run of whole model; large reduction in data requirements possible by aggregating (lumping) of subcatchments and channel/pipes, especially useful for continuous simulation. **(14)** Metric units optional for all input and output.

Ease of Application
(1) Nonproprietary model available from EPA, Athens, GA, or Department of Environmental Engineering Sciences, University of Florida, Gainesville, FL. (2) Updated user's manual and thorough documentation of most routines published as EPA reports; no one report covers all model aspects. (3) Test cases documented in several EPA and other reports. (4) Short course proceedings also useful for model applications. (5) US and Canadian user groups with newsletters and semiannual meetings permit publication of changes. (6) Due to its age (originally published in 1971), availability, and documentation, examples of SWMM usuage are widely available in the literature; bibliography available. (7) Frequent model update/corrections/improvements are often difficult to learn about; new model released approximately on a biannual basis. (8) Size of model most frequent deterrent to use; however, see item 13 under "Input Data Requirements" (preceding). (9) Initial model setup often moderately difficult due to size. (10) Model supported by EPA Center for Water Quality Modeling, Athens, GA.

Output and Output Format
(1) Input data summary including precipitation. (2) Hydrographs and pollutographs (concentrations and loads versus time) at any point in system on time-step or longer basis; no stages or velocities printed. (3) EXTRAN also outputs elevation of hydraulic grade line. (4) Surcharge volumes and required flow capacity; TRANSPORT Block will resize conduits to pass required flow (optional). (5) Stage, discharge, and soil moisture content for subsurface routing in RUNOFF Block. (6) Removal quantities in storage/treatment units, generated sludge quantities. (7) Summaries of volumes and pollutant loads for simulation period, continuity check, initial and final pounds of solids in conduit elements. (8) Daily (optional), monthly, annual, and total summaries for continuous simulation, plus ranking of 50 highest time-step precipitation, runoff, and pollutant values. (9) Line-printer plots of hyetographs, hydrographs, and pollutographs. (10) Costs of simulated storage/treatment options. (11) Statistical analysis of continuous (or single-event) output for event separation, frequency analysis, moments and identification of critical events.

Linkages to Other Models
(1) Linkage provided to EPA WASP and DYNHYD receiving-water-quality models. (2) Individual blocks and the total SWMM model have been linked to the HEC STORM model, the QUAL-II model, simplified receiving-water models, and others. (3) Individual blocks (e.g., RUNOFF Block) have been altered by various groups.

Personnel Requirements
(1) Civil/environmental engineer familiar with urban hydrological processes for data reduction and model analysis. (2) Computer programmer for model setup and establishment of off-line files on mainframes.

Costs
(1) Model and documentation available at little or no charge from EPA Center for Exposure Assessment Modeling, Athens, GA, 30613; also available for nominal charge from Department of Environmental Engineering Sciences, University of Florida, Gainesville, FL 32611. (2) Data assembly and preparation may require multiple man-weeks for a large catchment or urban area. (3) Example mainframe computer execution costs given in user's manual, on the order $20 for a RUNOFF and TRANSPORT run for a single-storm event with about 50 subcatchments and channel-pipes; use of EXTRAN can be more costly (greater than $100 per run) due to short time-step; continuous simulation (hourly time-step) of one subcatchment with snowmelt for two years costs about $20. (4) Microcomputer execution times depend on machine; execution times on AT-compatible approximately 50 times longer than for mainframe. (5) Extensive calibration may be required to duplicate measured quality results, quantity calibration relatively simple. (6) National Weather Service precipitation tapes for continuous simulation cost about $200 for at least a 25-yr hourly record for all stations in one state.

Model Accuracy
(1) Quantity simulation may be made quite accurate with relatively little calibration. (2) Quality simulation requires more extensive calibration using measured pollutant concentrations; quality results will almost certainly be very inaccurate without local measurements. (3) EXTRAN accurately simulates backwater, flow reversal, surcharging, pressure flow; TRANSPORT routines may be used at less cost if these conditions are not present. (4) Sensitivity to input parameters depends upon schematization; however, surface-quality predictions are most sensitive to pollutant loading rates.

mary is reproduced from the user's manual for the latest version of SWMM (Huber and Dickinson, 1988). This table was adapted from material appearing in publications by Huber and Heaney (1980, 1982). Beginning with the computational blocks, the fundamentals of each computational block and the more important service blocks are briefly described below. The user's manual should be consulted for more details.

RUNOFF Block

The RUNOFF Block generates surface runoff and pollutant loads in response to precipitation events and surface pollutant accumulations. The RUNOFF Block is a lumped model that is typically applied to a number of sub-catchments to distribute more accurately the physical properties of the watershed (recall the earlier discussion). In many ways, RUNOFF is the "heart" of SWMM.

The manner in which surface runoff is generated by RUNOFF is actually very simple and is summarized in Figure 2.12. Essentially, the surface is treated as a nonlinear reservoir that has a single inflow, precipitation. There are several outflows, including infiltration, evaporation,

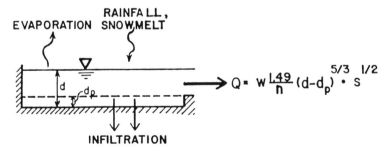

Figure 2.12. Nonlinear reservoir representation of a sub-watershed, RUNOFF block, SWMM (Huber and Dickinson, 1988).

and surface runoff. Infiltration occurs only if the ground surface is pervious (as opposed to an impervious surface, like a paved parking lot, which by definition allows no infiltration). The infiltration process is governed by one of two methods selected by the user (Horton's equation or the Green–Ampt equation). Infiltrated water is routed through upper and lower subsurface zones and may contribute to total runoff through groundwater flow.

Monthly average evaporation rates are provided by the user. These are used directly to lower the water in the "reservoir" and indirectly to calculate evaporation from the subsurface zones. Surface runoff occurs only when the depth of water in the "reservoir" exceeds the depth of depression storage. Depression storage refers to surface ponding, surface wetting, interception, and other storages that must be satisfied before surface runoff can occur.

Mathematically, surface runoff is generated by two equations. One is the continuity equation, which keeps track of the volume or depth of water on the surface (i.e., in the reservoir):

$$\frac{\delta V}{\delta t} = \frac{A \cdot \delta d}{\delta t} = (A \cdot i_e) - Q \tag{2-2}$$

where

V = $A \cdot d$ = volume of water on the sub-watershed, feet^3 or meters^3

A = area of the subwatershed, feet^2 or meters^2

d = depth of water on the subwatershed, feet or meters

t = time, seconds

i_e = rainfall excess, which is the rainfall intensity less the evaporation/infiltration rate, feet/second or meters/second

Q = runoff flow rate from the subwatershed, feet^3/second or meters^3/second

The differential term in equation 2-2 represents the change in water stored on the subwatershed per unit time. The first term to the right of the equal sign is the rainfall excess or net input rate to the subwatershed. The second term represents the discharge from the subwatershed, i.e., the runoff rate.

The second governing relationship is Manning's equation, which governs the rate of surface runoff (i.e., the discharge rate from the reservoir) as a function of the depth of flow above the depression storage level. For a very shallow "reservoir" Manning's equation can be written as:

$$Q = W \, (\beta/n) \, (d - d_p)^{5/3} \, S_o^{1/2} \qquad (2\text{-}3)$$

where
$\quad W$ = width of flow over the subwatershed, or the width of overland flow, feet or meters
$\quad \beta$ = 1.49 if U.S. customary units are used or 1.0 if metric units are used
$\quad n$ = Manning's roughness coefficient
$\quad d_p$ = depth of maximum depression storage, feet or meters
$\quad S_o$ = slope of the subwatershed, feet/foot or meters/meter

The full derivation and description of the conceptualization of the surface runoff process by equations 2-2 and 2-3 can be found in the SWMM user's manual (Huber and Dickinson, 1988) and a book by Nix (1994).

The most confusing aspect of RUNOFF is W or the "width of overland flow." Again, using the reservoir analogy, this width is similar to the length of a weir or spillway. An idealized view is shown in Figure 2-13. In this schematic, surface runoff is being discharged to a drainage

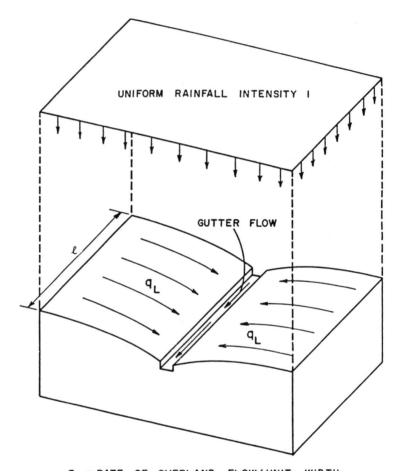

q_L = RATE OF OVERLAND FLOW/UNIT WIDTH.

W = 2ℓ = TOTAL WIDTH OF OVERLAND FLOW

Figure 2.13.   Idealized subwatershed–gutter arrangement illustrating the subwatershed width of overland flow, RUNOFF Block, SWMM (Huber and Dickinson, 1988).

channel located in the center of the subcatchment. If the two halves are symmetrical in all aspects, the total length of overland flow is twice the length of the channel. Of course, this idealized case never occurs, but it demonstrates the concept. The width of overland flow is not easily estimated from the physical characteristics of the watershed and is typically used as a "calibration" param- eter. This is especially true when adjusting the shape of the runoff hydrograph (the wider the subcatchment, the faster the water can be "discharged").

RUNOFF has the ability to route flows through gutters and pipes, using the nonlinear reservoir technique. How- ever, only three cross-sectional shapes are modeled and the routing technique is not adequate for many applica- tions. The sewer routing algorithms in the TRANSPORT and EXTRAN Blocks are more commonly used for this purpose. However, RUNOFF can be useful in modeling smaller lateral sewers, whereas TRANSPORT or EXTRAN focuses on the trunk sewers. The routing characteristics of these three blocks are discussed in the next section.

The surface flows generated by RUNOFF are concen- trated at nodes—i.e., the flows are not distributed along gutters or pipes (as implied by Figure 2.13). The width of overland flow is used as a computational tool but the flow is not actually distributed over this distance.

The accumulation of pollutants on the land surface is simulated in a number of ways. Pollutants can be accumu- lated or "built up" as "dust and dirt" on streets (i.e., pounds/100 feet of curb/day) or as a simple area-based load (e.g., pounds/acre/day). The rate at the which a pollutant accumulates may be linear or nonlinear. Each storm "washes off" the accumulated load according to the following "first-order" relationship:

$$P_{off} = \frac{-dP_p}{dt} = R_e \cdot r^m \cdot P_p \tag{2-4}$$

where

P_{off} = rate at which the pollutant is washed off at time t from the start of the storm, quan- tity/second

P_p = amount of pollutant p on the subwatershed surface at time t from start of the storm, quantity (the amount at the beginning of a storm comes from the surface accumulation algorithm)

r = runoff rate over the subwatershed at time t, feet/second or meters/second

R_c = washoff coefficient, (unit depth)$^{-m}$

m = exponent for the runoff rate

Equation 2-4 states that the rate at which a pollutant washes off the surface is proportional to the amount remaining on the surface and varies according to the intensity of the storm event. Alternatively, the user may specify that the washoff rate is strictly a function of the runoff rate (surface accumulation is ignored; this is essentially a rating equation relating the delivered pollutant load to runoff). The pollutant buildup and washoff routines can be confusing and the user should be careful. The data entry process is also very cumbersome.

There are many other capabilities not discussed here, most notably snowmelt simulation. The RUNOFF Block consumes a considerable portion of the SWMM user's manual, making it seem more profound and difficult than it is. Recall that the heart of the block is a very simple nonlinear reservoir representation of the surface runoff process.

TRANSPORT Block

The TRANSPORT Block routes flows and pollutant loads through a drainage or sewer system. These flows and loads may be generated by the RUNOFF Block or some other program and placed at points throughout the system. TRANSPORT also has the ability to generate dry-weather or sanitary sewage flows for routing through a sewer system. The user may also directly introduce constant flows or time-varying hydrographs as inflows to various points in the system.

The sewer system is viewed as a series of "elements" or "links and nodes" (see Figure 2.14). Elements may be nodes such as manholes, pump stations, and overflow structures or conduits linking the nodes. Conduits may have 15 different "regular" cross-sectional shapes (e.g., circular, rectangular) supplied by the program, an "irregular" or "natural" shape, or two supplied by the user. The different element types are listed in Table 2.3. The sewer

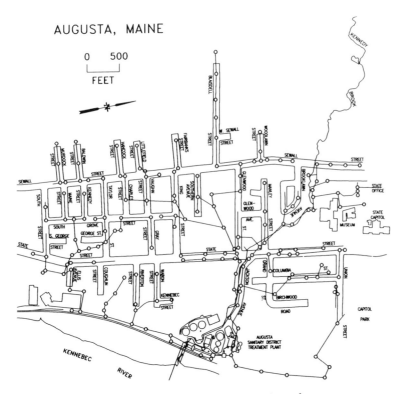

Figure 2.14. Node and conduit representative of a sewer system, TRANSPORT Block, SWMM (Heaney et al., 1975).

Table 2.3. Element types in TRANSPORT
Block, SWMM (Huber et al., 1988)

N Type	Conduit/Channels
1	Circular
2	Rectangular
3	Phillips standard egg-shaped
4	Boston horseshoe
5	Gothic
6	Catenary
7	Louisville semielliptic
8	Basket-handle
9	Semicircular
10	Modified basket-handle
11	Rectangular, triangular bottom
12	Rectangular, round bottom
13	Trapezoid
14	Parabolic
15	Power function
16	HEC-2 format—natural channel
17, 18	User-supplied
	NONCONDUITS
19	Manhole
20	Lift station
21	Flow divider
22	Storage unit
23	Flow divider—weir
24	Flow divider
25	Backwater element

system or network is specified by indicating which elements are upstream of each element. The elements are identified by user-supplied numbers.

Ideally, flow can be represented by two differential equations: the continuity and momentum equations or, as they are sometimes known, the Saint-Venant equations. A full description of these equations can be found in the original SWMM documentation (Metcalf and Eddy, Inc. et al., 1971) and the book by Nix (1994). Unfortunately, the Saint-Vevant equations are difficult to manipulate and simplifications are often desirable. TRANSPORT uses a simplified version of the momentum equation in which the dynamic effects are ignored and flow is just a function of depth. This simplification, known as a kinematic approximation, poses a severe restriction in that waves are propagated only in the downstream direction. Therefore, backwater effects, surcharging (the transition between open channel to pressure flow), looped connections, and pressure flows are not simulated or handled in very rudimentary ways. Probably the best way to understand how the model operates is to view the system as a cascade of conduits, with downstream conduits having no effect on upstream conduits. Simple representations of flow diversion devices (e.g., overflow structures) are also provided.

It is especially important to understand how the TRANSPORT Block behaves when it encounters surcharge conditions. Flows exceeding a conduit's full-flow capacity are stored at the upstream end of the conduit and released when the capacity again becomes available. Hydrographs passing through such a conduit become "clipped" and, as a result, the true downstream conditions are distorted. In short, if appreciable surcharge problems are suspected the TRANSPORT Block is not an appropriate model.

For comparison, the flow routing capabilities of the RUNOFF, TRANSPORT, and EXTRAN Blocks are summarized in Table 2.4. The RUNOFF Block, as discussed earlier, has a very limited routing algorithm. The TRANSPORT Block's abilities have just been outlined. The EXTRAN Block more completely represents the hydraulics of sewer systems and it is described in the next section.

Pollutants are routed very simply through the system

Table 2.4. Routing Capabilities of RUNOFF, TRANSPORT, and EXTRAN Blocks, SWMM (Huber et al., 1988)

		Blocks	
	RUNOFF	TRANSPORT	EXTRAN
1. Flow routing method	Nonlinear reservoir, cascade of conduits	Kinematic wave, cascade of conduits	Complete equations, interactive conduit networks
2. Relative computational expense for identical network schematizations	Low	Moderate	High
3. Attenuation of hydrograph peaks	Yes	Yes	Yes
4. Time displacement of hydrograph peaks	Weak	Yes	Yes
5. In-conduit storage	Yes	Yes	Yes
6. Backwater or downstream control effects	No	No*	Yes
7. Flow reversal	No	No	Yes
8. Surcharge	Weak	Weak	Yes
9. Pressure flow	No	No	Yes
10. Branching tree network	Yes	Yes	Yes
11. Network with looped connections	No	No	No

	3	16	8
12.	Number of preprogrammed conduit shapes	3	16	8
13.	Alternative hydraulic elements (e.g., pumps, weirs, regulators)	No	Yes	Yes
14.	Dry-weather flow and infiltration generation (base flow)	No	Yes	Yes
15.	Pollutograph routing	Yes	Yes	No
16.	Solids scour/deposition	No	Yes	No
17.	User input of hydrographs/pollutographs	No	Yes	Yes

*Backwater may be simulated as a horizontal water surface behind a storage element.

by treating each conduit as a completely mixed reactor with simple first-order decay. Solids scour and deposition may also be simulated.

In summary, working with TRANSPORT is relatively straightforward and easy. The mathematical algorithms are relatively stable. It is most effective with simple cascading sewer systems that are not subject to extensive surcharging or backwater conditions.

EXTRAN Block

The EXTRAN Block extends some of the capabilities of the TRANSPORT Block (although it has its own developmental history). This is accomplished by using more complete versions of the two Saint-Venant equations. Thus, it is capable of propagating waves upstream and, in turn, simulating backwater effects, surcharging, pressure flow, looped connections, etc. EXTRAN has acquired a reputation for being an unstable, difficult model that requires a fair amount of pampering. It is more difficult to operate than TRANSPORT but it definitely becomes less onerous with experience. The current version does not route pollutants, but this may change in the near future. A full description can be found in an auxilliary user's manual that accompanies the main SWMM documentation (Roesner et al., 1988), as well as the book by Nix (1994).

As in the TRANSPORT Block, the sewer system is viewed as a network of links and nodes (see Figure 2.14). The catalog of element types is not as extensive as that of the TRANSPORT Block (see Table 2.5) and the method of linking the system together is slightly different. Instead of specifying only the elements upstream of each element, upstream *and* downstream nodes are assigned to each link (or conduit). Inflows to the system occur at the nodes and

Table 2.5. Element Types in EXTRAN Block, SWMM
(Huber et al., 1988)

Element Class	Types
Conduits or links	Rectangular
	Circular
	Horseshoe
	Egg-shaped
	Basket-handle
	Trapezoid
	Power function
	Natural channel
	(irregular cross section)
Junctions or nodes (manholes)	—
Diversion structures	Orifices
	Transverse weirs
	Side-flow weirs
Pump stations	On-line or off-line pump station
Storage basins	On-line, enlarged pipes or
	tunnels
	On-line or off-line, arbitrary
	stage-area relationship
Outfall structures	Transverse weir with tide gate
	Transverse weir without tide
	gate
	Side-flow weir with tide gate
	Side-flow weir without tide
	gate
	Outfall with tide gate
	Free outfall without tide gate

may be entered directly by the user or come from other programs such as the RUNOFF Block.

The stability and accuracy of the numerical techniques used to solve the governing equations rely on the selection of an appropriate time step (recall the discussion early in this chapter). Fortunately, the user's manual gives clear instructions for estimating the proper values. Typically, the time step is less than 30 seconds. By comparison, the TRANSPORT Block can usually operate satisfactorily with time steps of many minutes. The short time step can result in relatively long run times, depending on the size of the drainage/sewer system. This and occasional stability problems are the most significant drawbacks of the EXTRAN Block.

Considerable effort is sometimes needed with EXTRAN, but its capability to model more-sophisticated drainage/sewer systems is often necessary. Four general rules can help:

1. Keep the system as simple as possible (while still meeting the modeling objectives).

2. Err on the side of using shorter time steps.

3. Remember that problems are most like to occur at nodes containing weirs, orifices, pumps, etc.

4. Inspect the output very carefully. Do not rely on summary tables.

STORAGE/TREATMENT Block

The STORAGE/TREATMENT Block is designed to route flow and pollutant loads through a storage and/or treatment facility. The user is given a great deal of flexibility by the block's ability to connect as many as five units together in a variety of networks. Each unit may be given detention (or storage) characteristics or be a simple flow-through device.

If a unit modeled as a detention unit flows are routed through the unit with the modified Puls method. This method requires information regarding the relationships between stage, storage, and discharge. Pollutants are routed by assuming that the unit behaves as a completely mixed reactor or a plug-flow reactor. Pollutant removal is accomplished through the use of removal equations or through discrete particle settling. In the former the program provides several state variables such as detention time, inflow rate, etc. around which the user can build a wide range of removal equations. In the latter method, each pollutant is removed based on its association with particles of various settling velocities, or sizes and specific gravities (e.g., 20% of the BOD load is associated with particles that have a given settling velocity).

When a unit is defined as a simple flow-through device, flow is routed without delay. Pollutant removal is simulated with a removal equation (again, built by the user with a number of state variables provided by the program).

The STORAGE/TREATMENT Block uses relatively simple representations of pollutant routing and removal processes. It is not intended to be used for the detailed simulation of unit operations. This block is intended to estimate how a storage/treatment facility will respond under dynamic stormwater flow conditions. As such, the modeling of pollutant routing and removal is fairly simple in order to keep the model tractable. Those wanting to simulate the nuances of a particular treatment technology will be sorely disappointed.

Receiving Water Modeling

Earlier versions of SWMM included an algorithm known as the RECEIVE Block, which was capable of routing flows and pollutant loads through receiving waters such as rivers, streams, and estuaries (Metcalf and Eddy, Inc. et al., 1971; Huber et al., 1975). The latest version of SWMM does not contain the RECEIVE Block. However, there are provisions in two EPA receiving water models,

DYNHYD4 and WASP4, for interfacing with SWMM (Ambrose et al., 1988).

Service Blocks

Some of the more important service blocks are described below.

The EXECUTIVE Block acts as the "control center" of SWMM. It directs the order in which the other blocks are executed and the passage of information from one block to another (and external programs). Interface files containing the information that must be transferred between blocks have a standardized format to ease the interaction between blocks and external programs.

The RAIN Block processes lengthy precipitation records for the RUNOFF Block. These records are usually read from National Weather Service tapes and diskettes, although user-generated records may also be processed. RAIN also incorporates an algorithm to perform storm event analyses. This routine is based on the well-known SYNOP program originally developed by Hydroscience, Inc. (1976, 1979).

The TEMP Block processes lengthy temperature and wind speed records for the RUNOFF Block. These records are usually read from National Weather Service tapes and diskettes, although user-generated records may also be processed.

The STATISTICS Block performs a limited number of statistical operations on a sequence of events (typically runoff events, although rainfall events may be processed as well). A variety of quantity and quality parameters may be analyzed. Output can include (at the user's option): a table showing the sequential series of events; a table of listing the magnitude, return period, and frequency of the events; graphs showing magnitude versus return period and frequency; and the first three moments of the event data. Events are most commonly separated and identified by a user-defined number of time steps of zero flow values (but other methods are available).

CONCLUDING REMARKS

This chapter presented a concise discussion of the use of mathematical models to simulate the stormwater in the municipal or industrial setting. Models are very helpful in understanding, projecting, and abating stormwater problems. But modeling must only be part of an integrated attack on a problem, not its focus. A model should substantiate more than shape the expert's judgment.

Much of the chapter focused on the use of the EPA Storm Water Management Model. This was not intended to minimize the potential of other models, but to provide a framework for discussion of how a typical model works. SWMM is a popular, well documented model, but it is not the only good stormwater model.

Fortunately, information about stormwater modeling is plentiful and new information appears regularly. There are several good sources on current theoretical developments and applications. A publication of the Water Environment Federation, *Water Environment Research*, produces an excellent annual literature review of the water pollution field, including a section dealing with "Urban Runoff and Combined Sewer Overflows" (see the June issue). The EPA Storm Water and Water Quality Model Users Group is informally sponsored by the U.S. Environmental Protection Agency and it meets about twice a year. Meetings are typically alternated between sites in the United States and in Canada. Proceedings are normally published. More information is available from the USEPA

Environmental Research Laboratory, College Station Road, Athens, Georgia 30613.

The final chapter in this book presents several case studies. All involve significant modeling efforts with SWMM. Case studies can provide useful insights but they are poor substitutes for a careful study of available models and for the diligent use of verification, validation, and sensitivity analysis tools to see that the chosen model performs credibly. Each situation is unique and demands conscientious analysis.

REFERENCES

Ambrose, R. B., S. B. Vandergrift, and T. A. Wool, *WASP4, A Hydrodynamic and Water Quality Model—Model Theory, User's Manual and Programmer's Guide,* EPA-600/3-87-039, U.S. Environmental Protection Agency, Athens, Georgia, 1988.

American Public Works Association, "Urban Stormwater Management," Special Report No. 49, APWA, Chicago, Illinois, 1981.

Brandstetter, A. B., "Assessment of Mathematical Models for Storm and Combined Sewer Management," EPA-600/2-76-175a, U.S. Environmental Protection Agency, Cincinnati, Ohio, August, 1977.

Chu, C. S., and C. E. Bowers, "Computer Programs in Water Resources," WRRC Bulletin 97, Water Resources Research Center, University of Minnesota, Minneapolis, Minnesota, November, 1977.

DiToro, D. M., and M. J. Small, "Stormwater Interception and Storage," *Journal of the Environmental Engineering Division,* ASCE, Vol. 105, No. EE1, February, 1979, pp. 43–54.

Donigian, A. S., Jr., and N. H. Crawford, "Modeling Pesticides and Nutrients on Agricultural Lands," EPA-600/2-76-

043, U.S. Environmental Protection Agency, Athens, Georgia, February, 1976.

Geiger, W. F., and H. R. Dorsch, "Quantity-Quality Simulation (QQS): A Detailed Continuous Planning Model for Urban Runoff Control, Volume I, Model Description, Testing and Applications," EPA-600/2-80-011, U.S. Environmental Protection Agency, Cincinnati, Ohio, March, 1980.

Hall, M. J., *Urban Hydrology*, Elsevier Applied Science Publishers, New York, 1984.

Heaney, J. P., W. C. Huber, and S. J. Nix, "Storm Water Management Model: Level I—Preliminary Screening Procedures," EPA-600/2-76-275, U.S. Environmental Protection Agency, Cincinnati, Ohio, October, 1976.

Heaney, J. P., and S. J. Nix, "Storm Water Management Model: Level I—Comparative Evaluation of Storage-Treatment and Other Management Practices," EPA-600/2-77-083, U.S. Environmental Protection Agency, Cincinnati, Ohio, 1977.

Howard, C. D. D., "Theory of Storage and Treatment-Plant Overflows," *Journal of the Environmental Engineering Division*, ASCE, Vol. 102, No. EE4, August, 1976, pp. 709–722.

Huber, W. C., "Deterministic Modeling of Urban Runoff Quality, *Urban Runoff Pollution*, H. Torno, J. Marsalek, and M. Desbordes, eds., NATO ASI Series, Series G: Ecological Sciences, Vol. 10, Springer-Verlag, New York, 1985, pp. 166–242.

Huber, W. C., "Modeling Urban Runoff Quality: State-of-the-Art," B. Urbonas and L. A. Roesner, eds., *Proceedings of Engineering Foundation Conference on Urban Runoff Quality—Impact and Quality Enhancement Technology*, ASCE, New York, June, 1986, pp. 34–48.

Huber, W. C., and Dickinson, R. E., "Storm Water Management Model, Version 4: User's Manual," Cooperative Agreement CR-811607, U.S. Environmental Protection Agency, Athens, Georgia, August, 1988.

Huber, W. C., and J. P. Heaney, "Operational Models for Stormwater Quality Management," M. R. Overcash and J. M. Davidson, eds., *Environmental Impact of Nonpoint Source Pollution*, Ann Arbor Science, Ann Arbor, Michigan, 1980, pp. 397–444.

Huber, W. C. and J. P. Heaney, "Analyzing Residuals Discharge and Generation from Urban and Non-Urban Land Surfaces," D. J. Basta and B. T. Bower, eds., *Analyzing Natural Systems, Analysis for Regional Residuals—Environmental Quality Management*, Resources for the Future, Washington, D.C., The Johns Hopkins University Press, Baltimore, Maryland, 1982, pp. 121–243. (This is also available as EPA-600/3-83-046, U.S. Environmental Protection Agency.)

Huber, W. C., J. P. Heaney, and B. A. Cunningham, "Storm Water Management Model (SWMM) Bibliography," EPA-600/3-85-077, U.S. Environmental Protection Agency, Athens, Georgia, September, 1985.

Huber, W. C., J. P. Heaney, M. A. Medina, W. A. Peltz, H. Sheikh, and G. F. Smith, "Storm Water Management Model User's Manual—Version II," EPA-670/2-75-017, U.S. Environmental Protection Agency, Cincinnati, Ohio, March, 1975.

Huber, W. C., J. P. Heaney, S. J. Nix, R. E. Dickinson, and D. J. Polmann, "Storm Water Management Model User's Manual, Version III," EPA-600/2-84-109a, U.S. Environmental Protection Agency, Cincinnati, Ohio, 1984.

Hydrologic Engineering Center, "Storage, Treatment, Overflow, Runoff Model, STORM," Generalized Computer Program 723-S8-17520, HEC, Corps of Engineers, Davis, California, August, 1977.

Hydroscience, Inc., "Areawide Assessment Procedures Manual," Vols. I, II, and III, EPA-600/9-76-014, U.S. Environmental Protection Agency, Cincinnati, Ohio, July, 1976.

Hydroscience, Inc., "A Statistical Method for the Assessment of Urban Stormwater," EPA-440/3-79-023, U.S. Environmental Protection Agency, Washington, D.C., May, 1979.

Johanson, R. C., J. C. Imhoff, and H. H. Davis, *User's Manual for Hydrological Simulation Program—Fortran (HSPF): Users Manual for Release 8.0*, EPA-600/3-84-066, U.S. Environmental Protection Agency, Athens, Georgia, 1984.

Metcalf and Eddy, Inc., University of Florida, and Water Resources Engineers, Inc., "Storm Water Management Model," Vols. I, II, III, and IV, EPA Reports 11024 DOC 07/71, 08/71, 09/71, 10/71, U.S. Enviromental Protection Agency, Washington, D.C., 1971.

Moffa, P. E., S. J. Nix, P. L. Freedman, J. Marr, and R. Field, *Control and Treatment of Combined-Sewer Overflows*, Van Nostrand Reinhold, New York, 1990.

Mueller, J. A., and D. M. DiToro, "Combined Sewer Overflow Characteristics from Treatment Plant Data," EPA-600/S2-83-049, U.S. Environmental Protection Agency, Cincinnati, Ohio, 1983.

Nix, S. J., "Applying Urban Runoff Models," *Water Environment and Technology*, WEF, Vol. 3, No. 6, June, 1991, pp. 47–49.

Nix, S. J., *Urban Stormwater Modeling and Simulation*, Lewis Publishers, Chelsea, Michigan, 1994.

Roesner, L. A., R. P. Shubinski, and J. A. Aldrich, "Storm Water Management Model User's Manual, Version III: Addendum I, EXTRAN," EPA-600/2-84-109b, U.S. Environmental Protection Agency, Cincinnati, Ohio, 1984.

Roesner, L. A., J. A. Aldrich, and R. E. Dickinson, "Storm Water Management Model User's Manual, Version 4: Addendum I, EXTRAN," Cooperative Agreement CR-811607, U.S. Environmental Protection Agency, Cincinnati, Ohio, August, 1988.

Sonnen, M. B., "Urban Runoff Quality: Information Needs,"

Journal of the Technical Councils, ASCE, Vol. 106, No. TC1, August, 1980, pp. 29–40.

Terstriep, M. L., and J. B. Stall, "The Illinois Urban Drainage Area Simulator, ILLUDAS," Bulletin 58, Illinois State Water Survey, Urbana, Illinois, 1974.

U.S. Environmental Protection Agency, "EPA Environmental Data Base and Model Directory," Information Clearinghouse (PM 211A), USEPA, Washington, D.C., July, 1983.

Whipple, W., Jr., N. S. Grigg, T. Grizzard, C. W. Randall, R. P. Shubinski, and L. S. Tucker, *Stormwater Management in Urbanizing Areas*, Prentice-Hall, Inc., Englewood Cliffs, New Jersey, 1983.

3

Receiving Water Impacts

INTRODUCTION

Protection of receiving water quality is the ultimate goal of stormwater control. As is the case with combined sewer overflow controls, it is expected that future stormwater regulations will require demonstration of compliance with water quality standards. To this end, any stormwater control program and the alternatives under consideration must be evaluated in terms of how they reduce receiving water problems and restore or maintain a beneficial use. The impact of stormwater loadings on water quality bears several similarities to, and a few distinct differences from, the impacts of combined sewer overflows (CSOs). The similarities are that both are intermittent loading sources driven by runoff events that deliver various pollutants from the watershed to the receiving water. In this regard, analysis of stormwater impacts is essentially identical to analysis of combined sewer overflows, as discussed previously in Moffa (1990). The primary differences relate to typical concentrations and number of sources. Stormwater inputs typically have lower concentrations than CSOs of those pollutants associated with sanitary waste (e.g., oxygen-demanding materials, pathogenic bacteria). Stormwater loading sources, however, are far more widespread, and have a greater number of loading sources. Figure 3.1 illustrates pollutants related to stormwater and the relative time scales.

This chapter will discuss the types of water quality

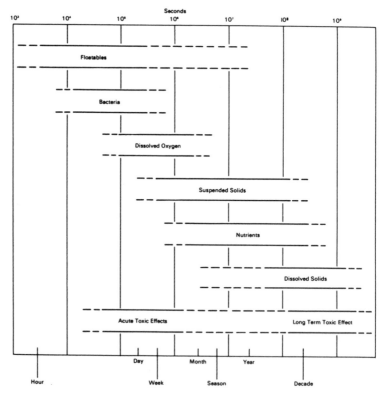

Figure 3.1.   Time scales for storm-runoff water-quality problems (EPA-625/4-79-013).

impacts that are associated with stormwater discharges, and how these impacts can be assessed as part of the stormwater control design process. The first section below, Water Quality Impacts, describes the problems associated with stormwater loading. The second section, Monitoring of Impacts, describes the collection of data required to identify existing problems and support the development of receiving water models. The third section, Modeling Approaches, describes the types of approaches that can be taken to model receiving water. The final section, Critical Design Conditions, describes the methods available for applying models to determine required stormwater loading controls.

WATER QUALITY IMPACTS

The specific receiving-water impacts from stormwater vary, depending upon the nature of the watershed, the quantity of runoff, the size and nature of the receiving water, and the designated use of the receiving water. For example, the nature of the watershed will determine whether runoff will contain agricultural pesticides or industrial by-products. The size and type of the receiving water body will determine its ability to assimilate stormwater discharges without impairment of the water's designed use. Therefore, receiving-water impacts must be evaluated and based on each site's individual characteristics. In each study, stormwater impacts (and benefits associated with reducing these impacts) can be evaluated with respect to three basic groups of consideration:

physical/chemical impacts

microbial impacts

aesthetic impacts

Each consideration affects the value of the receiving water for such uses as recreation, water supply, and ecological habitat. The significance of stormwater loads to the three major groups of consideration will vary, depending upon the nature of the load and the receiving water. Discussions regarding each type of impact are given below.

Physical/Chemical Impacts

The most common impacts of stormwater are measured in terms of changes in the concentration of certain chemical or physical constituents. In some cases, the significance of impairment can be directly determined via comparison to water quality standards (e.g., bacteria, toxics). In other cases, stormwater loads serve to initiate changes in other parameters (e.g., nutrients initiating changes in plant growth). These physical/chemical impacts can be divided into categories of:

nutrients

solids

oxygen-demanding materials

toxics

Nutrient enrichment is one of the most common causes of impairment associated with stormwater runoff. Excessive nutrients in the receiving water can cause problems by stimulating the growth of algae or rooted aquatic plants. Excessive plant growth can cause dissolved oxygen problems, reduce biologic diversity, worsen aesthetics, or impair use for water supply. Unfortunately, the response time from when nutrients are loaded to the water to when plant-related problems occur can be on the order of days or weeks (or longer). As a result, it is difficult to distinguish the impact of stormwater nutrient loadings from other sources over extended periods of time. Because of this, it is important to assess stormwater loadings in the context of overall nutrient loading to the waterbody. Stormwater nutrient loads may only contribute a small percentage of the total nutrient load. In these situations, stormwater controls will have little impact on overall water quality. In other cases, stormwater may be the dominant source of nutrients to a waterbody. In these cases, little improvement in water quality can be gained until stormwater sources are controlled.

Another problem associated with stormwater is the loading of excessive amounts of particulate material (solids) to the receiving water. This loading of solids can cause many receiving water problems. First, excessive receiving water solids concentrations can cause direct toxicity to aquatic organisms, through such mechanisms as fouling of gills, suffocation, etc. Second, high solids concentrations reduce water clarity, and can lead to violations of water quality standards for turbidity. Third, excessive solids loadings can settle out of the water column and cause various problems in the bedded sediments. Problems associated with sedimentation of particulate material is typically associated with relatively slow-moving waterbodies, as fast-moving streams possess the capability to scour deposited particulate material from the bottom. These sedimentation problems can be caused by both the physical and chemical properties of the sediments. The most common physical impairment to the bedded sediments caused by excessive solids loading is habitat destruction, where fine-grained sediments blanket ecologically important (e.g., spawning) areas. The most commonly observed type of chemical contamination of bedded sediments relates to the loading of particulate organic carbon.

This organic carbon material is manifested as a sediment oxygen demand, as the organic matter undergoes microbial degradation in the bottom sediments. Another potential sediment-related concern of stormwater loadings is the accumulation of toxics in the bottom sediments. The United States Environmental Protection Agency is currently considering promulgating Sediment Quality Criteria (SQC) that will limit the amount of toxic organics and heavy metals in the bottom sediments to "acceptable" levels. The intent of the SQC will be to maintain sediment contaminant concentrations such that interstitial pore water concentrations will be at or below applicable water quality criteria for chronic toxicity. The control of toxics in the bottom sediments should be assessed like nutrient control, because it consists of an integration over time of all loading sources. The significance of stormwater loading can only be determined by comparing the overall magnitude of stormwater loads to that of all other loading sources.

A third potential physical/chemical water quality problem associated with stormwater is dissolved oxygen depletion. Short-term reductions in dissolved oxygen due to BOD deoxygenation and ammonia nitrification can last from hours to days. Generally, dissolved-oxygen problems are more commonly associated with combined sewers (because of their higher BOD and ammonia concentrations). The increase in receiving water dilution that occurs during wet weather conditions is typically sufficient to counteract the increase in stormwater loading of oxygen-demanding materials. Dissolved oxygen problems associated with stormwater loading may occur, however, in weakly flushed waterbodies that receive relatively large quantities of stormwater loading.

Another water quality impact from stormwater loading is receiving-water toxicity. In general, toxicity problems fall into two categories: (1) acute toxicity, an increase in organism mortality due to shorter exposure to high toxicant concentrations; and (2) chronic toxicity, a decrease in organism viability due to long-term exposure to relatively lower toxicant concentrations. Due to the intermittent nature of stormwater loads, acute toxicity is generally more of a concern than chronic toxicity. Toxic discharges from stormwater are most commonly found in urban areas, where elevated concentrations of metals and some organics can be found. When assessing potential toxicity impacts of metals associated with stormwater, the bioavailability of metals in the receiving water should also be considered. Typically, only the dissolved, free ionic, form of heavy metals is responsible for toxicity. The high concentrations of suspended solids typically associated with stormwater tend to bind with most of the available metals, making them relatively nontoxic. Although these bound metals are not immediately bioavailable when they enter the water column, they may eventually desorb in the water column (or sediments) and become bioavailable. For this reason, some states choose to ignore the issue of metals bioavailability and assume that all metals entering the receiving water are bioavailable. Other states take a more rigorous approach of assessing the fraction of metals loading that will become available to cause toxicity. Depending upon the approach taken, metals loading reductions required to maintain compliance with water quality standards will vary substantially.

Microbial Impacts

Another source of concern related to stormwater is public health risks associated with diseases caused by pathogenic bacteria and viruses. Stormwater often carries excrement that can include these pathogenic organisms. Human exposure to stormwater-contaminated receiving waters therefore poses a risk of exposure to these pathogens, and is a concern in all types of receiving water. Levels of bacteria associated with stormwater are typically much lower than those found in combined sewer overflows; nonetheless, they are often several orders of magnitude greater than applicable receiving water quality standards. For this reason, stormwater loading of pathogens can cause violations of water quality standards in receiving waters even with a high degree of dilution. The risk is typically greatest at the very beginning of a stormwater loading event, due to the "first flush" effect that delivers much of the waste that had accumulated during the dry weather period. The risk can diminish quickly thereafter due to bacterial die-off, settling, and dilution. Significant public health risks rarely persist more than two or three days after a storm event. They are of special concern during warm weather when exposure is greatest. However, bacterial death rates are high during warm weather, tending to reduce exposure times.

The protection of public health from exposure to microbial pathogens is complicated by the extreme difficulty in measuring the concentration (or even the presence) of the pathogenic organisms. For this reason, water quality standards are based on the presence of more easily measured indicator organisms. The most commonly user indicator of fecal materials has been fecal coliform (FC) bacteria, based upon work of the U.S. Department of Interior's National Technical Advisory Committee in 1968. Many state water quality standards are based upon this work, and consist of a standard of 200 FC/100 ml for total body contact recreation, and 1000 FC/100 ml for partial body contact recreation. EPA revised these criteria in 1986 to be based upon concentration of *Escherechia coli* (a bacteria of the fecal coliform group) and enterococci (a subgroup of the fecal streptococcus group). Several states have adopted these new criteria and other states are expected to follow.

In general, stormwater loadings of pathogenic organisms will provide much less of a human health threat than will combined sewer overflows. Combined sewer overflows contain waste primarily of human origin, whereas stormwater contains waste primarily of animal origin. EPA's 1986 study concluded that the ultimate source of the primary agent for public health problems is human fecal material. As a result, states using the older criteria may show violations of the water quality standard for bacteria caused by stormwater, even with little risk to public health. The newer criteria will presumably be more accurate in representing the causative factors of public health impacts, and should result in fewer cases in which stormwater is falsely implicated for impairing a water's designed use.

Aesthetic Impacts

Aside from physical–chemical contamination and public health risks, stormwater can also impair the aesthetics of receiving waters. Stormwater can contain an assortment of unsightly materials, including floatables and oil and grease. Although aesthetic impairment may not reduce aquatic health or cause human health risk, it does have important socioeconomic significance. Resultant stormwater debris and suspended solids can reduce recreational use and appreciation of receiving waters, thereby limiting recreational expenditures and near-shore development. Because aesthetics are typically an integral component of a water's designed use, aesthetic impairment can be considered a violation of water quality standards. No good quantitative measures of aesthetic reduction are available, but visual observations and resident interviews often adequately define the problem. Alternatives can then be evaluated in terms of how effectively they control solids and floatables.

MONITORING OF IMPACTS

In every stormwater study, there will be a debate on the best means to define impacts and benefits for alternative control strategies. Should it be monitoring or modeling? Practically speaking, it is not an either/or decision, but rather a question of how much of each. The two activities need to be balanced and designed for an integrated assessment. Monitoring is needed to identify existing problems and to support the development of reliable and verified models. Nothing provides stronger evidence of receiving-water impacts than actual measurements and observations. Therefore, all stormwater evaluation studies should include measurements and observations of post-storm receiving-water impacts due to stormwater loading. However, monitoring has its limitations. The monitored conditions

1. may be unrepresentative

2. may not capture infrequent but severe events

3. do not delineate the separate effects of stormwater and other nonpoint and point source pollution; and

4. cannot be used to compare proposed stormwater controls.

Therefore, modeling should generally be used to complement monitoring and to test and analyze conditions other than those monitored. On the other hand, models need to be supported by actual data; therefore, monitoring and modeling must be designed together as an integrated program with full consideration of the types and extent of data needed to define the problems. (More will be provided on modeling in later sections of this chapter.)

Monitoring programs should include one or more initial wet-weather reconnaissance surveys involving a few stations downstream of stormwater inputs and at least one station upstream. The reconnaissance monitoring should be conducted during seasons expected to encompass the most significant impacts, and extend from a day before a storm to as much as four days afterwards. Parameters should include BOD, dissolved oxygen, temperature, bacteria, and other parameters suspected as having impacts, plus visual observations for debris and floatables. Occasionally, routine fixed-frequency monitoring data can be used to assess the presence of impacts and the periods or seasons when the receiving water is most susceptible to stormwater discharges. However, most often the timing of the fixed-frequency samplings does not coincide well with storm impacts. Therefore, surveys must be designed with this specific need targeted.

The reconnaissance survey(s) should be used as a basis to design more intensive surveys in terms of sampling location, duration, and frequency. Subsequent to the reconnaissance survey(s), at least two or three additional intensive surveys are recommended to better define impacts for different conditions and support model development. Often, impacts are evident only for large storms and are not apparent for smaller events. Hence, multiple surveys are recommended for defining the recurrence of problems and representing a fuller range of conditions.

In general, station locations and frequency for intensive surveys must be in sufficient numbers to define spatial and temporal trends clearly. This usually requires sampling every 3–6 hours at stations located with time of travel between locations of approximately 6–12 hours. In rivers, the sampling logistics are the simplest because all impacts flow downstream. Station locations and timing are easy to define. For estuaries, station location is still relatively easy, but consideration of tidal movement and timing becomes important. Samplings at fixed times in each tidal cycle (usually a slack) are most desired to help standardize presentation and analysis of the data. For lakes, timing is less difficult, but station location becomes more difficult to choose due to multidimensional transport. In such cases, drogue or current studies can be used to define current transport. However, lake circulation patterns can vary depending on wind and other factors. Therefore, a network of lake stations should be set up to encompass the probable horizontal and vertical transport directions. Preliminary data can be used to refine the sampling network. Also, reconnaissance data may show minimal localized effects, in which case coarse whole-lake monitoring may be more appropriate than a localized network. In all cases for rivers, estuaries, and lakes, the reconnaissance sampling should be used as a guide to establish station locations and frequency.

The sampling parameters should focus on the anticipated problems. However, all stormwater impact surveys should routinely include measurement of dissolved oxygen, BOD, ammonia, and fecal coliform bacteria. Oftentimes, available site-specific data are available to define which parameters will be of concern. Temperature and visual observations of cloud cover, sunlight, water transparency, floatables, debris, and other conditions are inexpensive and should also be collected routinely. Daily composites are usually suitable for other parameters such as nutrients and chlorophyll. Toxic substances, such as trace metals, typically need only be measured immediately before the storm to define a baseline, immediately after the storm to define maximum impacts, and at the end of the survey to determine persistence.

Some special note should be made concerning the scheduling of stormwater impact monitoring. Obviously, it is impossible to "plan" to have storms occur conveniently for sampling. It is also generally impractical to use automatic samplers for the receiving water. It is, therefore, very important to have a well-developed mobilization plan in order to capture immediate post-storm impacts. Impacts related to solids, bacteria, and toxics are best captured immediately after the storm, during the first flush of pollutants from the sewers. A rapid mobilization is essential. The use of private weather forecasting is often a valuable aid. Also, routine fixed-frequency monitoring programs can be a valuable source of pre-storm data and identifying accumulative impacts.

With data in hand, the planner/engineer can proceed to analyze more fully a wider range of conditions, using a mathematical model. Models are highly suited for providing information not available from data alone. In particular, models can be used to simulate water quality for conditions not monitored. This could include larger, more intense storms, or more-critical environmental conditions, such as lower stream flow. It is generally infeasible to extrapolate to nonmonitored conditions from data, due to the large number of variable factors. The power of modeling is that all such conditions can be assessed quickly. Furthermore, models can be used to examine impacts and benefits from proposed stormwater control strategies.

In overview, monitoring is essential to all stormwater studies to demonstrate and define the presence of stormwater impacts. Modeling provides an additional tool to analyze impacts at nonmonitored conditions. However, models must be based on and confirmed with actual physical, chemical, and biological information. Therefore, monitoring and modeling must be conducted hand-in-hand.

MODELING APPROACHES

Mathematical modeling can be a powerful tool for assessing stormwater impacts and for comparing stormwater control alternatives. However, the effective use of modeling requires the selection of an appropriate framework and then site-specific validation with data. General guidance is provided in this chapter for the selection and development of suitable models for stormwater assessments. Mathematical models use algebraic and differential equations to simulate the physical, chemical, and biological processes that determine water quality. Modeling consists of three basic components:

1. model formulation
   a. system equations
   b. equation solutions
2. model validation
   a. model inputs
   b. comparison to data
   c. coefficient adjustment
3. model use
   a. selection of conditions
   b. comparison of results

Several factors, which include technical as well as management issues, must be considered in the definition of an appropriate modeling framework. Selection of an appropriate model framework requires integration of:

site characteristics

pollutant characteristics

management objectives

project constraints

The significance of each factor is discussed briefly immediately below, with specific practical application described later in this section. The most significant site characteristic influencing selection of a receiving water model is the degree of spatial gradients in water quality. Only a very few, complex, models can consider two- or three-dimensional gradients in water quality, whereas many models are available for zero- or one-dimensional problems. The pollutant(s) being considered will also dictate to a large extent the required modeling framework. Two totally separate classes of receiving water models exist to consider conventional pollutant impacts (e.g., dissolved oxygen depletion) as compared to toxic pollutant impacts. The specific management objectives of a modeling effort also play a role in selecting a modeling framework. Typically, management objectives can be condensed into two equations: (1) What specific questions do we want this model to answer?; and (2) How reliable must the predictions be? The answers to these questions will dictate the degree of complexity required in the model selected. Certain project constraints may dictate an upper bound to the level of modeling complexity that should be attempted. The four primary constraints are:

(1) schedule, (2) budget, (3) data to support modeling efforts, and (4) technical resources (staff expertise, computers). As a rule, no model should be selected whose requirements exceed any of these four constraints. Provided an appropriate model has been chosen, the next step is to define inputs such as flow, pollutant loads, temperature, and reaction rates to characterize the site properly. Then, before use in problem solving, the model reliability should be validated by comparison to data. Overall, despite the apparent burden of these steps, models are essential tools in analyzing conditions and alternatives not monitored. Only through the use of model simulation can the engineer/planner compare the merits of different stormwater control strategies.

Helpful Concepts

The first question usually asked of models is: "What is the impact of stormwater on water quality?" Fortunately, water-quality models are very well suited for analyzing the incremental impacts from various sources. The nature of the equations used in water-quality models allows the application of the concept of superposition. This mathematical concept allows each individual source of pollution impact to be modeled separately. The total water-quality condition is then the sum of component impacts simulated separately.

Superposition is a powerful technique that allows the modeler flexibility in answering important questions. For example, stormwater impacts can be incrementally compared to other point source impacts or to baseline water quality. Establishing the relative importance of stormwater impacts to other sources can be very important in developing cost-effective, basin-wide pollution solutions, and establishing the need for and benefit from stormwater controls. Superposition can also be used to compare an individual stormwater loading source to help devise cost-effective prioritization of controls.

Analyzing stormwater impacts separately from other sources is an important conceptual approach. Stormwater control programs are often based on an evaluation of incremental improvements, not absolute compliance to water-quality standards. This approach is often preferred because wet-weather conditions are too variable and factors other than stormwater often control compliance to water-quality standards.

A common dilemma among new modelers is how to conveniently analyze the transient-nature stormwater impacts. The first approach is often to utilize complex models that can characterize all the transient aspects of the system and storm under study. However, two simpler approaches are available. One can always assume that the loading and related conditions are constant or steady-state. This provides a conservative estimate of the impact, but often is representative if loading points are close together and dispersional mixing is small. Analyzing the problem in a moving reference (Lagrangian) framework is another approach. This involves simulating the dynamics of a plug of polluted waters as it flows or circulates downstream, ignoring the dynamics and surrounding system. This approach is very useful if stormwater loading points are either very close together and can be analyzed together, or are far enough apart that they can be analyzed separately.

Having provided some overall conceptual guidance on modeling apparatus, some specifics related to different types of receiving-water bodies can be examined. Again, the objective of this chapter is general guidance, not textbook instruction on modeling. For more-thorough technical guidance, the reader is referred elsewhere (e.g., Thomann and Mueller, 1987).

Streams and Rivers

The modeling mechanics for streams and rivers are the most simple because the dilution flow is well quantified and the direction of flow obvious. If one assumes that the stormwater pollutant inputs are well mixed across the stream/river channels, then the characteristic equations are simple. The simple dilution/first-order loss equation is a good example. This classic equation (and related dilution equations) can be used to simply characterize water quality impacts in many streams and river. The basic equation is:

$$C = C_0 \exp(-Kt)$$

where

C_0 = initial pollution concentration (M/L^3)
C = pollutant concentration downstream (M/L^3)
W = total pollutant loading (M/T)
Q = total river flow (L^3/T)
t = time of passage (T)
k = pollutant decay rate $(1/T)$.

By using the concept of superposition, this basic equation can quickly and simply be used to characterize stormwater impacts in many streams and rivers. Additional pollutant sources or components can be analyzed separately and are added to baseline conditions to simulate water quality.

The apparent complexities in stormwater analysis, due to transient stormwater loading, are easily handled in streams and rivers. The modeler can either conservatively assume that all the loads are steady and compound, or instead focus attention on only one plug of water moving downstream. This allows an assumption of constant loading (average for the plug) and then a tracking of this plug as it goes downstream. Higher and lower loading rates can each be analyzed separately because the pollutants travel in a train with limited intermixing. Additional compounded impacts from downstream loads can be added through superposition by separately tracking the time of passage with the time of the loadings.

The preceding equation is obviously very basic, and the reader is again referred to other standard modeling textbooks for more details and added complexities (Thomann and Mueller, 1987). Nonetheless, these basic equations and approaches are common to all modeling of stormwater impacts. Various computer models are available that utilize these equations and concepts as a basis, but offer added flexibility to handle complexities in reactions, loadings flow, and geometry.

Estuaries

Stormwater impacts on estuaries can often be addressed in a manner similar to that used for stream systems, but with additional complexity due to dispersion, tidal hydraulics, and saline intrusion. If the upper freshwater portion of an estuary is being examined, then simple one-dimensional, intratidal analyses can be appropriate. If assessments require examination of intertidal, short-term impacts or involve the lower portion of the estuary (where stratification due to saline intrusion is a major process), then more-complex two- and three-dimensional dynamic computer models may be needed.

Impacts of stormwater discharge in the upper estuary can often be analyzed in terms of a slug of impacted water, as for streams and rivers. However, in addition to the advective transport in rivers, the slug is also affected in estuaries by tidal motion and dispersional mixing. This tends to distribute and dilute the slug over time. In addition, because tidal action moves the slug back and forth as it progresses downstream, modeling must either account for this motion or be calculated as consistent points in the repeating tidal cycle. Surveys used for model development are often conducted at the points of low or high slack tides to allow comparison to intratidal modeling results.

As in streams and rivers, there are analytical, closed-form equations to simulate the transport, mixing, and reaction of pollutants in estuaries. The calculations conservatively assume the loadings are steady, or assume the loadings occur as a instantaneous slug. In both cases, the equations can be used to calculate the transport downstream, the mixing due to tides, and the reaction gains or losses.

The slug of pulse loading equation is usually preferred for analyzing stormwater impacts because the steady-state equations underestimate the dilution effects from tidal mixing. Assuming full lateral and vertical mixing, the basic water quality equations to simulate the impacts of pulse loading are:

$$C = SM\exp(-Kt)$$
$$S = \exp[-(x - ut)^2/4Et]/[2A(pEt)^{1/2}]$$

where

A = area
E = longitudinal dispersion (L^2/T)
x = distance downstream (L)
M = mass loading of a pollutant (M)
U = average velocity of stream (L/T)

and other terms are as applied before. Again, the impact of other sources can be added by superposition.

The approach just shown allows quick calculation of stormwater impacts in estuaries on a tidal-average basis. This approach generally requires the modeler to sum nearby loads and ignore small-scale differences due to intertidal processes. Rapid and large tidal mixing generally lends greater validity to these assumptions. Therefore, these simple equations provide powerful methods for screening alternatives and comparing stormwater impacts.

Although the desktop formula approaches are highly recommended for many situations, site and problem characteristics can often dictate the need for a more rigorous and complex analysis. Widely distant, varied, and oddly timed stormwater loads, coupled with highly variable estuary conditions, are examples of factors limiting the applicability of the simpler analysis. In these cases, the modeler must generally use one of several computer models specifically adapted for estuaries, such as WASP4 (EPA, 1988), and intertidal models. For estuary sites having multidimensional transport, the level of model complexity is beyond the scope of this chapter.

Lake Analysis

Analyzing the impacts of stormwater on lakes and impoundments is much different and more complex than for rivers and estuaries. First, and most obvious, the flow and circulation in lakes is much more varied and complex. Second, the nature of water quality problems and processes is often different in lakes. As a result, lake analysis usually requires specialized computer modeling programs and comprehensive monitoring programs. Stormwater impacts in lakes can generally be categorized into three groupings: immediate end-of-pipe impacts, whole-lake impacts, and spatially patterned impacts. End-of-pipe impacts and whole-lake impacts can often be analyzed using simple analytical equations, but spatially patterned impacts almost always require a complex computer model.

End-of-pipe impacts are those in the immediate discharge plume from a stormwater outfall and generally involve acute toxicity concerns. The research on plume mixing and dilution is extensive, and numerous formulae are available to calculate the impacts, depending on the physical characteristics of the outfall site (Fischer et al., 1979; EPA, 1985). Factors to consider when selecting an individual formula include

near-field versus far-field

outfall submergence

discharge buoyancy

discharge velocity

shore and bottom effects

The formulae are typically simple to apply, but proper selection is essential for reliable calculations.

Whole-lake impacts are very different and consider the accumulation of pollutants from stormwater on a lake-wide basis. Again, research in this area is extensive, and different formulae are available to address different types of problems for lakes with different characteristics (Chapra and Reckhow, 1983). In general, the simpler equations are based on a completely mixed characterization of lake with flow in and flow out. To calculate lake transients from a slug of pollutants completely mixed in the lake, the basic equation is:

$$C = C_0 \exp[(-K - Q/V)t]$$

where

$$
\begin{aligned}
Q &= \text{flow } (L^3/T) \\
V &= \text{lake volume } (L^3).
\end{aligned}
$$

and other terms are as defined previously. For a steady-state constant load calculation, the lake equation is:

$$C_{SS} = \frac{W}{(Q + KV)}$$

where

C_{SS} = steady-state pollutant concentration (M/L^3).

Both of these equations and their respective assumptions are simplistic. They are usually suited only for performing a first-screening analysis to define the potential of stormwater impacts in lakes. The first approach usually underestimates more-localized impacts, whereas the second approach may overestimate the impacts. Nonetheless, the calculations are valuable for developing a quantitative feel for lake response to stormwater inputs.

In contrast to end-of-pipe and whole-lake effects, spatially patterned impacts are much harder to analyze. These are transient impacts that originate from stormwater but circulate, disperse, and dissipate as waters move the impacts to different lake locations. To characterize these impacts, complex computer models are needed. These models rely on the same principles of mass conversation as the simpler models, but usually consist of a complex network of calculation cells. Mass balance equations are written for each cell and then solved simultaneously by a computer program.

Sediments

The impacts of stormwater on river, estuary, and lake sediments are a common concern but with no easy means for analysis. There are no simple calculations that can be performed to assess these impacts. Unfortunately, comprehensive sampling and modeling are typically required. The potential of a problem can often by determined by an examination of pollutant concentrations in stormwater solids, and a comparison of stormwater solids with other sources of solids. However, confirmation of the problem and evaluation of the benefits from stormwater controls require more-rigorous analysis. Discussion of these considerations is beyond the scope of this chapter.

CRITICAL DESIGN CONDITIONS

Traditional water quality planning analyses have focused on control of continuous point sources of pollutants, whose water quality impacts are most significant during low stream flow. Well-established procedures are in place for performing these analyses. The most common assessment method for determining allowable loads for continuous sources is the "critical conditions" approach. In this method, controls are required that will protect instream water quality during some "critical" period (typically drought upstream flow), under the assumption that these controls will be sufficient for most other periods. Water-quality-based controls have only recently been required for stormwater runoff, and procedures for determining the required levels of control are not well established.

The impact of stormwater on receiving water depends on dilution, dictated by flow, and stormwater loading rate and volume, dictated by rainfall. Selecting a combination of the conditions as a critical design condition is difficult because the two do not ordinarily vary together in a predictable fashion. High stormwater loads from large rains do not typically coincide with low drought flows. As a result, a variety of combinations can be critical, ranging from moderate rains at drought flow to torrential rains at moderate flows. Hence, selecting one appropriate critical condition is often impractical. Nonetheless, evaluating stormwater impacts and benefits of proposed controls requires that water quality be identified along with associated abatement costs. Two general approaches are feasible for these evaluations. First, analyses can be conducted over the entire range of meteorological and flow conditions that have occurred historically, using a continuous simulation approach. Alternatively, a set of design conditions can be specified to represent expected impacts during "critical" environmental conditions.

Continuous Simulation Analysis

The one method that can directly account for compliance with water quality standards when considering intermittent loads is continuous simulation analysis. With continuous simulation analysis, runoff loads and receiving water quality are simulated over a long period of historic record, and the number of the water quality standard exceedance is tabulated (see Figure 3.2). If exceedances occur more often than is allowable (i.e., once in three years, allowed by water quality standards), loading controls must be placed on the pollutant sources. The continuous simulation model can be used to test various control strategies to determine which ones will lead to compliance with water quality standards. The continuous simulation approach, although the only rigorous method to assess required wet-weather loading controls, is very resource (data and time) intensive.

Design Condition Approach

EPA water quality guidance recommends that water quality criteria be exceeded no more than once in three years.

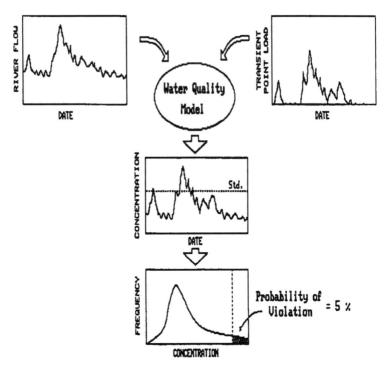

Figure 3.2.   Continuous simulation.

Designing controls for episodic wet-weather loading events has been problematic, because the degree of control required to meet this frequency objective is not easily defined. Dynamic modeling analyses (e.g., continuous simulation) allow direct consideration of the frequency of occurrence, but are considered too resource intensive for widespread application. A design condition approach, similar to that currently used for NPDES permitting of continuous discharges, is desired for episodic wet-weather loads.

Two factors have prevented development of wet-weather design conditions: (1) selection of an appropriate background streamflow; and (2) selection of an appropriate design storm. Selection of an appropriate stream flow has been problematic because of the correlation between stream flow and rainfall. A worst-case approach of assuming that a design rain event occurs during drought stream flow conditions may lead to extremely overprotective design. As noted in the WEF Manual on Urban Runoff, the joint probability of occurrence (assuming independent events) of a once-in-ten-year low flow and a once-in-one-year storm is once in 3,650 years. An approach proposed in the WEF Manual is to back-calculate a receiving water design flow that, when combined with a selected rainfall/runoff frequency, results in a probability of occurrence (once in 10 years) that coincides with the precedent that has been established for dry-weather point source discharges. Given a once-in-one-year design storm, and assuming that rainfall and stream flow are independent, the method suggests that a receiving water flow that is exceeded 328.5 days of the year is appropriate for design storm analysis. This method is more appropriate than the worst-case approach described above, but it is still quite simplistic in ignoring the correlation between rainfall and receiving water flow.

An alternative approach, intermediate in complexity between simplistic design storm analyses and continuous simulation, is described below and is patterned after EPA's DESCON model. DESCON analyzes the entire period of record for the inputs required for a simple wasteload allocation model (e.g., upstream flow, upstream pollutant concentration), and selects an assimilative capacity with a once-in-three-year return period. The existing DESCON allows a relatively precise estimate of design conditions, as the simple dilution model can be algebraically rearranged to provide an exact solution for assimilative capacity. The wet-weather DESCON cannot be as precise, as the relationship between rainfall and NPS load cannot be easily defined *a priori*. Fortunately, the wet-weather DESCON approach described herein requires only an estimate of relative NPS load magnitude, to allow a ranking of storm severity.

The proposed approach is summarized in Figure 3.3.

① Calculate Severity

Date	Flow	Rainfall	Severity $\left(\frac{W(\alpha\,Rain)}{Q}\right)$
1/1/60	385	0	0
1/2/60	497	1.8	0.004
1/3/60	678	0	0
.	.	.	.
.	.	.	.
.	.	.	.
12/28/92	315	0	0
12/29/92	164	0	0
12/30/92	27	0.9	0.033
12/31/92	201	0	0

⇓ RANK

② Determine 1/3 Year Design Rain/Flow

Severity	Flow	Rainfall	Return Period
0.033	27	0.9	6 yrs
0.004	497	1.8	3 yrs
.	.	.	.
.	.	.	.
.	.	.	.
0	385	0	0
0	678	0	0
0	315	0	0
0	164	0	0
0	201	0	0

③ Determine Design Intensity Pattern

Figure 3.3.   EPA DESCON model summary.

Specifically, the historical record of site-specific rainfall and stream flow data is analyzed, and ranked in terms of potential severity of wet-weather impacts. The appropriate storm/stream flow pair will be selected as the one that represents the once-in-three-year return frequency.

The approach requires the assumption that the magnitude of the non-point source load can be directly correlated from the observed rainfall. However, because this assumption is used only to rank relative storm severity (not to predict absolute loads), the final results should not be overly sensitive to this assumption.

The first two steps provide a method for pairing appropriate storms and stream flows, but do not address selection of specific storm characteristics. One option is to use the characteristics of the specific storm that was picked as part of the design flow/rainfall volume pairing. This will be protective for most sites, but allows the possibility that lower-volume/higher-intensity storms may lead to standards' being violated more frequently than once in three years. A safety factor can be added to the process, as shown in Step 3 of the figure, where the 95th (or 99th) percentile storm intensity pattern is paired along with the already selected stream flow and storm volume. This is a direct corollary to the approach described in EPA's "Technical Support Document for Water Quality-Based Toxics Control" of setting Long-Term Average NPDES effluent concentration limits such that the wasteload allocation concentration is exceeded only 1% (or 5%) of the time. Once the specific design storm characteristics are specified, they can be used by the permittee as input to the non-point source model of choice for use in designing appropriate controls.

References

Chapra and Reckow, 1983. *Engineering Approaches for Lake Management*, Vol. 2: *Mechanistic Modeling*, London: Butterworths.

Environmental Protection Agency, 1988. *WASP4, A Hydrodynamic and Water Quality Model—Model Theory, User's Manual, and Programmer's Guide*, EPA 600 3-87 039.

Fischer et al., 1979. *Mixing in Inland and Coastal Waters*, Academic Press.

Moffa, Peter E., 1989. *Control and Treatment of Combined Sewer Overflows*, New York: Van Nostrand Reinhold.

Thoman, R. V., and Mueller, J. A., 1987. *Principles of Surface Water Quality Modeling and Control*, New York: Harper & Row.

4

Stormwater Pollution Abatement Technologies

INTRODUCTION

This chapter covers the control and treatment of stormwater in relation to the removal or reduction of the stormwater pollutant loads. The prevention of flooding by the control of stormwater is not the emphasis of this chapter, however. Although many of the pollution abatement technologies discussed will help attenuate stormwater flows, the technologies will not provide sufficient capacity for large storm events, because they are generally designed for small events. Although prevention of stormwater flooding is not discussed in this chapter, a drainage system design should consider both pollutant and flooding aspects of stormwater.

GENERAL APPROACH AND STRATEGY

Small Storm Hydrology

The selection of suitable abatement technologies requires an understanding of the size and distribution of storm events. These factors contribute to total volume of storm runoff and, with knowledge of the pollutant concentrations, provide the total pollutant load. Generally the smaller storm events are the critical storms to consider, because for many parts of the U.S., 85% of all the rains are less than 0.6 in. (15 mm) in depth but can generate about 70% of the total annual storm runoff (Pitt, 1987). The characteristics of small and large storm events can be very different in terms of the storm runoff generated, pollutant load, and receiving water impacts.

However, the frequent small storms will have a more persistent impact; the less frequent large storms will have a larger impact, but they allow time for recovery between events. For small storm events, any inaccuracy in the estimation of the initial abstractions and the soil infiltration rates can significantly change the calculated storm runoff pollutant load. The initial abstractions include the rainfall depth required to satisfy surface wetting, surface depression storage, interception by hanging vegetation, and evaporation. Together with soil infiltration rates, the initial abstractions need to be accurately estimated to allow calculation of the storm runoff volume. Initial abstractions for relatively impervious urban surfaces have been found to account for the first 0.2–0.4 in. (5–10 mm) of a storm event (Pitt, 1987). Other researchers (Pecher, 1969, Viesman et al., 1977) have reported initial abstractions of between 0.02 and 0.14 in. (0.5–3.5 mm) for pavement areas, depending on whether the areas are flat or sloping steeply.

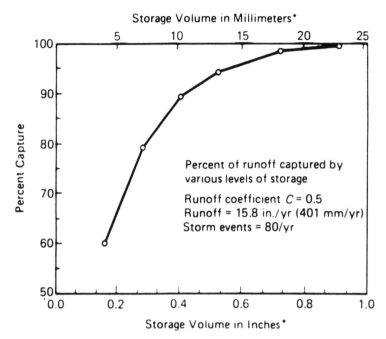

Figure 4.1.   Runoff capture volume rates in Cincinnati, Ohio
(Urbonas and Stahre, 1993).

Figure 4.1 illustrates the runoff capture volume rates in Cincinnati, Ohio. Note that 95% of the runoff will be captured for the first 0.5 watershed in. (12.7 mm) (recall that 85% of all storms are less than 0.6 in., or 15 mm). This indicates that small precipitation events need to be considered when designing stormwater quality treatment facilities. Increases in design detention volume above this limit will not significantly affect the percent capture (Urbonas and Stahre, 1993).

Traditional stormwater flood control is concerned with the peak storm runoff flow rates from relatively infrequent large storm events and their conveyance. This is a different set of criteria to those needed for storm runoff pollution control. Therefore, the use of data, storm runoff coefficients, models, etc., intended or developed to meet stormwater flood control requirements should be used with caution. This is illustrated by initial abstractions that can be a major portion of a small storm but will be a relatively insignificant portion of a large storm. In other words, just because a model for an area has been verified as providing accurate information for large storm events does not mean it will predict small events with the same level of confidence.

It should not be taken from the above that the large infrequent storm events do not cause polluted urban storm runoff or significant impacts on receiving waters but that their infrequency makes them a less significant factor than the smaller frequent storms.

There are several other factors that will affect the stormwater runoff pollutants and their concentrations, as discussed elsewhere, and these will also need to be taken into consideration when estimates are made of the urban storm runoff pollutant load.

A model developed and currently being updated for the calculation of urban stormwater runoff pollutant loads from small storms is *Source Loading and Management Model: An Urban Nonpoint Source Water Quality Model (SLAMM)* (Pitt, 1989). This model concentrates on the parameters discussed above to better estimate the urban storm runoff pollutant loads before and after application of best management practices (BMPs). However it is mainly applicable to small areas and does not give a continuous time analysis. There are, however, a number of other models, such as the U.S. EPA's Storm Water Management Model (SWMM), that will allow a continuous time analysis for large drainage areas. Continuous time analysis will provide an optimum design for storage and treatment facilities based on long-term historical weather patterns.

Strategy

The intermittent, widespread, and variable nature of urban stormwater runoff will require a flexible and creative approach to achieve the optimal control and treatment solution. This approach is likely to incorporate regulations, Best Management Practices (BMPs), and treatment processes. Traditional wastewater treatment methods, particularly secondary treatment processes that tend to operate under conditions closer to steady state, will not necessarily be suitable for the fluctuating loads of stormwater runoff. On the other hand, technologies used to control and treat combined sewer overflows (CSOs) are more likely to be applicable for the stormwater runoff, and advantage should be taken of any experience or facilities of CSO origin that have application for separate stormwater runoff. Successful stormwater management to control urban storm runoff pollution will require an area-wide approach combining prevention, reduction, and treatment practices/technologies. It is highly unlikely that any single method will provide the best solution to control the widespread diffuse nature of stormwater runoff and achieve the water quality required.

Establishing an urban storm runoff pollution prevention and control plan requires a structured strategy that will include the following steps:

define existing conditions

set site-specific goals

collect and analyze data

refine site-specific goals

assess and rank problems

screen BMPs and treatment technologies

select BMPs and treatment technologies

implement plan

monitor and re-evaluate

It is very likely that advantage can be taken of previous studies for either stormwater or CSO to get a head start. The above strategy is described in *Handbook: Urban Runoff Pollution Prevention and Control Planning* (U.S. Environmental Protection Agency, 1993a). Additional references that describe planning approaches for urban storm runoff pollution prevention and control are contained in Table 4.1.

The above strategy will provide the control goal to be achieved. The goal is then used as the basis for selection of suitable technologies or approaches. The goals should initially be broad and not specific, because the process of reviewing the technologies or appoaches available will in itself generate information to focus and refine the goals to meet cost, level of control, public opinion, feasibility, and other restraints. A flexible approach that through an iterative process of review and adjustment is focused upon a specific action plan, is the only real method by which the complexity of urban stormwater can be managed. The specific action plan will also need to be subject to reassessment once feedback on its implementation is available.

The above is only a very brief indication of the extensive work that will be required before the actual abatement technologies are implemented, and more detail is given in the reference U.S. Environmental Protection Agency, 1993a. The remainder of this chapter is concerned with an overview of the abatement technologies available. The chapter reviews the technologies by separating the drainage system into three physical areas:

watershed area (i.e., storm runoff generation/collection area)

installed and/or modified/natural drainage system (i.e., conveyance pipes, channels, storage, etc.)

end-of-pipe (i.e., point source)

Technologies applicable to each of these areas are discussed and can be divided into structural and nonstructural approaches. The nonstructural ones will cover approaches such as public education and regulations that will have their main application to the upstream collection area. The structural approaches will be the main options for the drainage system and end-of-pipe areas, and tend to be the more expensive items.

The technologies and approaches for stormwater management referred to as BMPs generally cover the nonstructural or low-structural stormwater runoff controls. The point at which a stormwater management technology changes from a BMP to a unit treatment process (i.e., "high-structural" control) is often unclear. Therefore, in this chapter, *BMPs* refer to the upstream watershed area prevention and/or control measures only.

As stated previously the optimal solution is likely to be an integrated approach using several practices and technologies. The management of the watershed using BMPs to prevent or control pollution at the source is likely to offer the most cost-effective solution and tends to be the basis of many stormwater management plans. However, although BMPs will be the preferred option they will not always be feasible or by themselves sufficient to achieve the control objectives. For older and more heavily urbanized areas BMPs are likely to have a limited application, and some form of treatment prior to discharge may be required. There are number of publications cited in Table

Table 4.1. Planning Approaches Suggested in Various Literature References

	Urban Surface Water Management (Walesh, 1989)	*Developing the Watershed Plan* (U.S. EPA, 1991a)	*Developing Goals for Nonpoint Source Water Quality Projects* (U.S. EPA, 1991b)	*Santa Clara Valley Nonpoint Source Study—Volume II: NPS Control Program* (SCVWD, 1990)	*State of California Storm Water Best Management Practice Handbooks* (Camp Dresser and McKee, 1993)	*Urban Storm Water Management and Technology: Update and User's Guide* (U.S. EPA, 1977a)
Determining existing conditions	Establish objectives and standards Conduct inventory	Identify problems and opportunities and determine objectives Develop resource data	Inventory resources and forecast conditions	Initiate public participation Define existing conditions Review regulatory problems Define goals and objectives	Define goals Assess existing conditions	Assess existing data Compare conditions vs. objectives Determine extent of runoff problem
Quantifying pollution sources and effects	Analyze data and prepare forecasts	Interpret, analyze, and evaluate data and forecasts	Identify problems Develop goals or objectives	Define and describe problems	Set priorities	Conduct selective field monitoring Refine problem estimates

	Formulate alternatives Compare alternatives and select recommended plan	Formulate and evaluate alternatives Evaluate and compare alternatives	Formulate alternatives Evaluate alternatives	Identify NPS control measures Evaluate control measures Develop evaluation criteria Examine and screen measures Select measures	Select near-term BMPs	Assess alternatives
Assessing alternatives						
Developing and implementing the recommended plan	Prepare plan implementation program Implement plan	Select alternative and record decision	Select best alternatives and record decision	Recommend control measures and implementation program	Implement near-term program Assess program effectiveness	Determine attainable improvements

Table 4.2. Urban Runoff and CSO BMP References

Document Title	Author or Editor	BMPs Included	Information Available
Controlling Urban Runoff: A Practical Manual for Planning and Designing Urban BMPs, 1987	Schueler	Detention Infiltration Vegetative Filtration Quality inlets	General description Effectiveness Design Use limitations Maintenance Cost Examples
Protecting Water Quality in Urban Areas, 1989	MPCA	Housekeeping Detention Infiltration Vegetative Quality inlets	General description Effectiveness Use limitations Maintenance Cost Examples
Guide to NPS Control, 1987	U.S. EPA	Housekeeping Detention Infiltration	General description Effectiveness Cost
Water Resource Protection Technology: A Handbook of Measures to Protect Water Resources in Land Development, 1981	Urban Land Institute	Housekeeping Detention Infiltration Vegetative Quality inlets	General description Effectiveness Design Use limitations Maintenance Costs

Reference	Source	Categories	Description items
Urban Targeting and Urban BMP Selection, 1990	Woodward–Clyde	Housekeeping, Detention, Infiltration, Vegetative	General description, Effectiveness, Design, Use limitations
Combined Sewer Overflow Pollution Abatement, 1989	WPCF	Housekeeping, Collection system, Storage, Treatment	General description, Design, Effectiveness, Maintenance, Cost
Urban Stormwater Management and Technology: An Assessment, 1974	U.S. EPA	Housekeeping, Collection system, Storage, Treatment	General description, Design, Maintenance, Use limitations
Decision Maker's Storm Water Handbook: A Primer, 1992	Phillips–U.S. EPA Region V	Housekeeping, Detention, Infiltration, Vegetative, Filtration, Quality inlets	General description, Effectiveness, Design, Use limitations, Maintenance, Examples
Urban Storm Water Management and Technology: Update and User Guide, 1977	U.S. EPA	Source control, Collection system, Storage, Treatment	General description, Design, Maintenance, Use limitations
Control and Treatment of Combined Sewer Overflows, 1993	Moffa	Source Control, Collection system, Storage, Treatment	General description, Design, Maintenance, Use limitations

Table 4.2. Urban Runoff and CSO BMP References

Document Title	Author or Editor	BMPs Included	Information Available
Coastal Nonpoint Source Control Program: Management Measures Guidance, 1993	U.S. EPA	Housekeeping Infiltration Vegetative Filtration Quality inlets	General description Effectiveness Design Use limitations Maintenance Cost Examples
The Florida Development Manual: A Guide to Sound Land and Water Management, 1992	Livingston et al	Housekeeping Infiltration Vegetative Quality inlets	General description Effectiveness Design Use limitations Maintenance Cost Examples
Storm Water Management Manual for the Puget Sound Basin, 1991	WA DOE	Housekeeping Infiltration Vegetative Quality inlets	General description Effectiveness Design Use limitations Maintenance Cost Examples

Stormwater Management, 1992	Wanielista and Yousef	Water quality Infiltration Detention	General description Effectiveness Examples Cost
Stormwater: Best Management Practices and Detention for Water Quality, Drainage, and CSO Management, 1993	Urbonas and Stahre	Storage Source control Detention Treatment Water quality	General description Effectiveness Design Use limitations
Integrated Stormwater Management, 1993	Field, O'Shea, and Chin	Detention Management Vegetative Infiltration Flood control Reclamation Collection systems	General description Effectiveness Design Use limitations

4.2 that cover the present state of the art on stormwater management using BMPs but that do not generally review the end-of-pipe treatments that could be applied to stormwater as a final line of control. This chapter will therefore place greater emphasis on the treatment options that are available for stormwater pollution control and that appear to be ignored in many stormwater management manuals.

It should, however, be emphasized that it will be more cost effective to prevent potential urban storm runoff pollution problems and protect existing resources than to construct pollution controls once a problem exists. Unfortunately, for many areas the problems already exist and retrospective prevention is not a feasible solution.

The implementation of any stormwater management program will need to meet financial and probably schedule restraints; therefore, an early review and improved utilization of existing facilities can offer several advantages. These options are likely to be the quickest and least costly to be implemented but it is also important they meet the objectives developed from the earlier stormwater management planning process. Examples include the enforcement of existing regulations to control soil erosion during construction activities and adaptation of existing stormwater storage intended for flood control to also provide quality control for small storm events. New installations should consider design for both flood control and pollutant removals.

The public does not generally perceive stormwater to be an environmental pollution problem. Furthermore, they do not appreciate the direct connection between some of their actions and the pollution consequences (e.g., disposal of engine oil and household toxic liquids down a storm drain or throwing litter that is transported by storm runoff into the receiving water). Gaining public support to cooperate in the implementation and pay for a stormwater management plan will be a major challenge. A strategy of concentrating efforts and resources on high-priority areas where results are likely to be achieved, and be seen to be achieved, will help generate public support.

WATERSHED AREA TECHNOLOGIES AND PRACTICES

There are many BMPs, but not all BMPs are suitable in every situation. It is important to understand which BMPs are suitable for the site conditions and can also achieve the required goals. This will assist in the realistic evaluation, for each practice, of the technical feasibility, implementation costs, and long-term maintenance requirements and costs. It is also important to appreciate that the reliability and performance of many BMPs have not been well established, with most BMPs still in the development stage. This is not to say that BMPs cannot be effective, but that they do not have a large bank of historical data on which to base design to be confident that the performance criteria will be met under the local conditions. The most promising and best understood BMPs are detention and extended-detention basins and ponds. Less reliable in terms of predicting performance, but showing promise, are sand filter beds, wetlands, and infiltration basins (Roesner et al., 1989).

A study of 11 types of water quality and quantity BMPs in use in Prince George's County, Maryland (Metropolitan Washington Council of Governments, 1992a) was conducted to examine their performance and longevity. The report concluded that several of the BMPs had either failed or were not satisfying the designed performance. Generally wet ponds, artificial marshes, sand filters, and infiltration trenches achieved moderate to high levels of removal for both particulate and soluble pollutants. Only wet ponds and artificial marshes demonstrated an ability to function for a relatively long time without routine maintenance. BMPs that were found to perform poorly were infiltration basins, porous pavement, grass filters, swales, smaller "pocket" wetlands, extended-detention dry ponds, and oil/grit separators. Infiltration BMPs had high failure rates, which could often be attributed to poor initial site selection and/or lack of proper maintenance.

The report cited above contains many more details and recommendations on the use of BMPs. It is important to note that the reported poor performance of some of the BMPs is likely to be a function of one or more factors: the design, installation, maintenance, and suitability of the area. Greater attention to these details is likely to reduce significantly the failure rate of BMPs. Other important design considerations include: safety for maintenance access and operations; hazards to the general public involving safety (e.g., drowning) or nuisance (e.g., mosquito breeding area); acceptance by the public (e.g., enhancement of area aesthetics); and assumption of conservative performances in the design until the historical data can justify a higher reliable performance.

For any BMP involving soil infiltraton of the storm runoff it is important to consider the possible effects this process could have on the ground water. These could range from a relatively minor local raising of the water table, with reduced infiltration rates, to more serious pollution of the ground water, particularly if it is also used as a water source. Stormwater runoff is likely to have very low levels of pollution when compared to chemical and gasoline leaks/discharges, and the soil will have some natural capacity to hold pollutants. However, the long-term build up of pollutants in the soil and/or groundwater from storm runoff infiltration is not well known. Therefore infiltration of urban storm runoff, particularly from industrial and commercial areas that are likely to have higher levels of pollution, should be treated with caution.

Infiltration of storm runoff can offer significant advantages in controlling storm runoff at the source, reduced risk of downstream flooding, recharge of groundwater, and groundwater supply to streams (e.g., low-flow augmentation or maintenance of stream flow during dry-weather periods). All of these and probably other advantages can be offered at a relatively low cost by infiltration, and therefore the advantages will need to be judged against any pollution risks from urban runoff.

The majority of treatment processes that can be readily applied to urban storm runoff are effective only for removal of the settleable solids. Removal of dissolved or colloidal pollutants will be minimal and therefore pollution prevention or control at the source offers an effective way to control the dissolved pollutants. Fortunately, however, many pollutants in the form of heavy metals and organic chemicals show significant association with the suspended solids (SS) (Pitt and Field, 1990, Pitt et al., 1991, 1993). Therefore, removal of the solids will also remove the associated pollutants.

The previously mentioned goals for a stormwater management plan can be achieved in the watershed area via three basic avenues.

Regulations and public education. This should be the primary objective because it is likely to be the most cost effective. Mainly nonstructural practices will be involved and application to new developments should be particularly effective.

Source control of pollutants. This will be closely related to the above. Both nonstructural and structural practices can be used to prevent pollutants from coming into contact with the stormwater and hence being present in storm runoff. Management and structural practices will include: flow diversion practices that keep uncontaminated stormwater from contacting contaminated surfaces or keep contaminated stormwater from contacting uncontaminated stormwater by a variety of structural means; exposure minimization practices that minimize the possibility of stormwater contacting pollutants by structural (diking, curbs, etc.) and management (coverings, loading and unloading) practices; mitigative practices, which include plans to recover released or spilled pollutants in the event of a release; preventative practices, which include a variety of monitoring techniques intended to prevent releases; control of sediment and erosion by vegetative and structual means; and infiltration practices, which provide for infiltration of stormwater into the groundwater (structural and vegetative means), thereby reducing the total runoff.

Source treatment, flow attenuation, and storm runoff infiltration. These are mainly structural practices to provide upstream pollutant removal at the source, controlled stormwater release to the downstream conveyance system, and ground infiltration or reuse of the stormwater. Upstream pollutant removal provides treatment of stormwater runoff at the location where it enters the stormwater conveyance system. Areas of this type will drain to a drainage system and include, but are not limited to, vehicular parking areas, vehicular service stations, bus depots, and industrial loading areas.

The following section provides brief details of BMPs, but the reader should appreciate the fact that many of these BMPs can be combined and/or modified to best suit the conditions of the watershed under consideration.

More information on BMPs can be found in the references listed in Table 4.2.

Regulations and Public Education

The regulatory approach can address a wide variety of stormwater management aspects, some of which are listed below. For any regulations to work, there will need to be an existing framework within which to place the regulations (e.g., ordinances, zoning, planning regulations), together with dedicated resources to enforce the regulations. Without the institutional systems to set them up and enforce them, they will not be effective.

Regulations can be an important pollution prevention BMP, with particular application to new developments to ensure that the pollution is prevented or controlled at the source and any implementation and maintenance costs are included in the development costs. New York State has compiled a manual on BMPs for new developments (New York State, 1992).

Some typical regulations include:

Land use regulations

- zoning ordinances
- subdivision regulations
- site plan review procedures
- natural resource protection

Comprehensive storm runoff control regulations

Land acquisition

Further details on a regulatory approach are contained in *Handbook: Urban Runoff Pollution Prevention and Control Planning* (U.S. Environmental Protection Agency, 1993a) and *Urban Stormwater Management and Technology: Update and Users' Guide* (U.S. Environmental Protection Agency, 1977).

Public education can have a significant role to play because an aroused and concerned public has the power to alter behavior at all levels. However, if the stormwater management plans are not adequately communicated and public opinion responded to, this power of the public can work against the implementation of a stormwater plan if the plan is viewed as an unnecessary extra cost and restriction on freedom.

Gaining the public support, as with all education, does not stop but is a continuous process and applies to all sectors of the public. These sectors are listed below and discussed in the following paragraphs:

residential

commercial

industrial

governmental

The residential sector is a made up of everyone living in a drainage area and therefore education should focus on large groups. Long-range education goals can be tackled through school programs and shorter-range goals may be achieved through community groups. Advantage should be taken of working with groups looking for community improvement projects and opportunities arising from news media coverage and the associated publicity.

The commercial sector is a fairly large and often diffuse group to communicate with. Both the owners/managers and their staff will need to be included in any communication, together with new businesses opening; existing businesses moving, expanding, and closing; and employee turnover. Methods of communication include news announcements in the local press, mailed news items, individual contact by a public official, and followup repeated contacts to answer questions and cope with employee turnover. Public education can benefit from failures, such as violations of regulations that result in a citation or fine and are reported in the local press. This not only informs the public about regulations but also provides an incentive for the regulations to be followed because they are seen to be enforced.

The industrial sector is a smaller group and can be educated by direct contact with public officials, through education of the consultants from whom industry seeks advice, and through education of trade associations. Indirect education opportunities are provided by speaking to meetings of professional organizations and by writing in professional newsletters and journals. Industrial decision makers are a relatively small group, which when informed or made aware of their obligations are likely to respond.

Public officials should also communicate with other public officials and governmental institutions to ensure that they are aware of stormwater management programs and their implications. Examples include road, sanitation, and parks departments, and workers at public institutions such as hospitals and prisons.

A multilevel, multitarget public education program can help to avoid problems in implementing a stormwater management program. Further information on communicating a stormwater management program to the public can be found in *Designing an Effective Communication Program: A Blueprint for Success* (U.S. Environmental Protection Agency, 1992a), and *Urban Runoff Management Information/Education Products* (U.S. Environmental Protection

Agency, 1993b). The latter reference is a catalog of material and publications that are available.

Source Control of Pollutants

Source controls are generally nonstructural practices, many of which can be termed "good housekeeping" practices. They can be very effective in that they are pollution prevention options. Some of them include:

cross-connection identification and removal

controlled construction activities

street sweeping

solid waste management

animal waste removal

toxic and hazardous waste management

reduced use of fertilizer, pesticide, and herbicide

reduced roadway sanding and salting

material and chemical substitution

Research on illicit or inappropriate cross-connections into separate stormwater drainage systems has shown that these can add a significant pollutant loading (Montoya, 1987; Pitt and McLean, 1986; Schmidt et al., 1986; and Washtenaw Co., 1988). This fact is also recognized in the National Pollution Discharge Elimination Permits System (NPDES) for stormwater discharges, which requires investigation of dry-weather flows (DWF) at stormwater outfalls. This will involve inspecting outfalls for DWFs, identifying illicit discharges from analysis of DWF samples, tracing the discharge source, and carrying out corrective action. DWF can originate from many sources. The most important sources may include sanitary wastewater (from sewer lines or septic tank systems), industrial and commercial pollutant entries, and vehicle maintenance activities. It should be recognized that not all DWF will be a pollutant source and may be caused by infiltrating potable water supply and clean groundwater. A full illicit-connections investigation is likely to be time consuming and costly. A methodology for identifying illicit discharges in the DWF and tracing the source using distinct characteristics of potential sources is described in *Investigation of Inappropriate Pollutant Entries into Storm Drainage Systems: A User's Guide* (U.S. Environmental Protection Agency, 1993c) and *Investigation of Dry-Weather Pollutant Entries into Storm Drainage Systems* (Field et al., 1994). The *User's Guide* concentrates on procedures that are relatively simple and that do not require sophisticated equipment or training. At a minimum the procedures should identify the most severely contaminated outfalls to assist in prioritizing areas to be investigated first, and at best will identify the pollutant source. A stormwater management plan that ignores investigation of DWF is very likely to find that goals set to improve receiving water quality will not be achieved due to pollutants discharged in DWF.

Soil erosion from construction sites together with wash-off from stockpiled material and ready-mix-concrete trucks can be a major source of pollutants (suspended solids) for the relatively short construction duration. Requirements for phased removal of vegetative cover and early reestablishment of ground cover, combined with detention of stormwater for sedimentation and filtering, will help reduce the pollution from construction site stormwater runoff. It is important also to consider the period following construction when vegetative ground cover has still to be fully established and occupants of new buildings may undertake landscaping. Further information can be found in *Reducing the Impacts of Stormwater Runoff from New Development* (New York State, 1992) and *Storm Water Pollution Prevention for Construction Activities* (U.S. Environmental Protection Agency, 1992b).

Street-sweeping studies (U.S. Environmental Protection Agency, 1979d, 1985) concluded that: typical reduction in storm runoff pollutant loadings can be between 5 and 10% for street sweeping carried out every two days (sweeping more than two days a week does not significantly reduce the solids loading any further, as illustrated in Figure 4.2); street cleaners did not significantly remove the smallest particles (<100 μ) that the rain washes off;

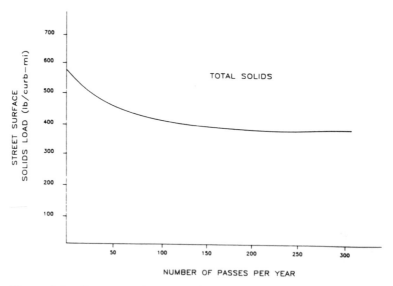

Figure 4.2.   Street sweeping: annual amount removed as a function of the number of passes per year at San José test site (U.S. EPA, 1979d).

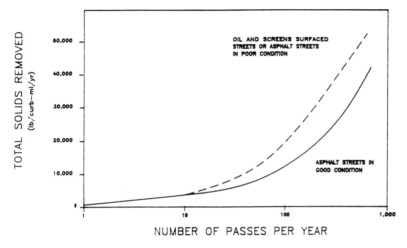

Figure 4.3. Street cleaner productivity in Bellevue, Washington (U.S. EPA, 1985).

street cleaners were able to remove large fractions of large particles (>200 μ); the reduction in storm runoff pollutant load is much less than the pollutant load removed by sweeping, which can lead to a false sense of effectiveness; pavement type and condition affect performance more than do differences in equipment performance (Figure 4-3 illustrates the variable nature of the total solids removed versus the number of passes per year in relation to pavement condition); and street sweeping results are highly variable and the results from one city cannot be applied to another city. The above comments, together with the fact that street storm runoff is only a part of the outfall discharge, imply that street cleaning is not particularly effective on its own but should be part of an overall program. Street cleaning is likely to be more effective for removal of heavy metals from vehicle emissions that tend to associate with the particulate. Sweeping of parking areas, storage, and loading/transfer areas should be included in a cleaning program. Concentrated cleaning during certain seasons is likely to be effective—e.g., during early spring in the snowbelt, when leaves accumulate in the fall, and prior to rainy seasons. Although the effectiveness of the above is not generally proven, street cleaning does offer aesthetic improvements in the removal of large items from the streets and receiving water. Fugitive emissions from street sweeping will lead to increased air pollution and may need to be considered if an intensive street-sweeping program is part of a stormwater management plan.

Solid waste management involves the collection and proper disposal of solid waste to maintain clean streets, residences, and businesses. It can also be extended for collection of items such as leaves during the fall. A study of stormwater runoff into Minneapolis lakes found that phosphorus levels were reduced by 30–40% when street gutters were kept free of leaves and lawn clippings (Minnesota Pollution Control Agency, 1989).

Domesticated and wild animal wastes represent a source of bacteria and other pollutants such as nitrogen that can be washed into the receiving waters. A study in San Francisco, California (Colt 1977) estimated that the dogs, cats, and pigeons produced 54,500, 9,000, and 2,200 lb (247, 4, and 1 metric tonnes), respectively, of nitrogen a year. On an annual basis, bulk precipitation, dog wastes, and fertilizer, respectively, accounted for 49, 23, and 22% of the total nitrogen runoff. Controls through regulation and public education, if successful, could therefore have a major impact based on these figures.

Toxic and hazardous waste management should review methods to prevent the dumping of household and automotive toxic and hazardous wastes into municipal stormwater inlets, catchbasins, and other storm drainage system entry points. Public education, special collection days for toxic materials, and posting of labels on stormwater inlets to warn of the pollution problems of dumping wastes are possible management options.

Fertilizers, pesticides, and herbicides washed off the ground during storms can contribute to water pollution. Agriculture, recreation parks, and gardens can be sources of these pollutants. Controlling the use of these chemicals on municipal lands, and educating gardeners and farmers to use the minimum amounts required and to use appropriate application methods can help reduce nutrient and toxic pollutants washed off by storm runoff.

Sand and salt are applied as deicing agents to roads in many areas of the United States that experience freezing conditions and are then washed off by the meltwater and stormwater runoff. Effects of highway deicing appear most significant in causing contamination and damage of groundwater, public water supplies, roadside wells, farm supply ponds, and roadside soils, vegetation, and trees (U.S. Environmental Protection Agency, 1971). Deicers also contribute to deterioration of highway structures and pavements, and to accelerated corrosion of vehicles. Studies (U.S. Environmental Protection Agency, 1971) indicated that major problems in the control of deicing chemicals were excessive application, misdirected spreading, poor storage practices, inaccurate weather forecasting, and the logistics of setting up the deicing operation. To address these problems two manuals of practice on the application and storage of deicing chemicals (U.S. Environmental Protection Agency, 1974a, 1974c) were produced to give recommendations and suggest improvements. They provide comprehensive details on storage management, layout, handling, application for various storm and temperature conditions, and use and calibration of equipment to minimize the amount of chemicals used. Studies were conducted on alternative deicing methods (U.S. Environmental Protection Agency, 1972a,

1976a, 1978) but these were more costly than the use of rock salt and therefore are unlikely to have general economic application.

SOURCE TREATMENT, FLOW ATTENUATION, AND STORM RUNOFF INFILTRATION

Vegetative BMPs

These developing practices have been the subject of many recent publications, a few of which are listed in Table 4.2. Readers are directed to these and similar publications for more-detailed information.

Knowledge of the performance of these systems and practices is limited, but the above publications do contain lessons learned from their implementation and, in some cases, failure. Existing urbanized areas are unlikely to have the land space available for installation of many of these practices and in these situations their application will be restricted.

Swales

These are generally grassed stormwater conveyance channels that remove pollutants by filtration through the grass and infiltration through the soil. A slow velocity of flow, <1.5 ft/s (<46 cm/s); nearly flat longitudinal slope, <5%; and vertical stand of dense vegetation higher than the water surface, ≥6 in. (15 cm) total height, are important for effective operation (Metropolitan Washington Council of Governments, 1992b). Swales can be enhanced by the addition of check dams and wide depressions to increase storm runoff storage and promote greater settling of pollutants. A further enhancement would be the development of a wetland channel (Urbonas and Stahre, 1993), but good design would be necessary to minimize the disadvantages of difficult maintenance access, mosquito breeding, and unpleasant aesthetics, to maximize the benefits of greater treatment potential.

Filter Strips

These are vegetated strips of land that act as "buffers" by accepting storm runoff as overland sheet flow from upstream developments and providing treatment potential mechanisms similar to those of swales, prior to discharge of the storm runoff to the storm drainage system. Low-velocity flows, installation of a level spreader and/or land grading to ensure sheet flow over the filter strip, and dense vegetative cover will enhance the filter strip performance (Metropolitan Washington Council of Governments, 1992b, Yu et al., 1993).

Stormwater Wetlands

These can be natural, modified natural, or constructed wetlands that remove pollutants through sedimentation, plant uptake, microbial decomposition, sorption, filtration, and exchange capacity. It is important to note that natural wetlands will be covered by regulations that limit what can be discharged to the wetland and any modifications to enhance the wetland performance. Constructed stormwater wetlands can be designed for more effective pollutant removal with elements such as: a forebay for solids capture; meandering flow for extended detention of low flows; benching of bottom for different water depths and associated plants; and pondscaping with multiple species of wetland trees, shrubs, and plants (Metropolitan Washington Council of Governments, 1992b).

Constructed wetland systems are increasingly being used and developed for wastewater treatment, and this area could be a source of information (e.g., Water Environment Federation, 1990, U.S. Environmental Protection

Agency, 1988) in addition to information in stormwater
BMP publications.

Detention Facilities

One of the most common structural controls for urban
storm runoff and pollution loading is the construction of
local ponds (including wetlands) to collect storm runoff,
hold it long enough to improve its quality, and release it
to receiving waters in a controlled manner. The basic
removal mechanism for detention ponds is through settling of the solids with any associated pollutants, but
controlled release will also attenuate the stormwater
flows, which can be a benefit to receiving water streams
that suffer from erosion and disturbance of aquatic habitat
during peak flow conditions.

It should be realized that a detention facility designed
to provide pollution control for a particular size of storm
is not likely to provide the same level of treatment for
smaller or larger storms. For example, a detention facility
designed to capture and release over a certain period a
10-year storm event may need to have the discharge control orifice designed for a 2-year storm in addition to the
10-year storm to provide discharge control and hence
treatment over a spread of storm events (Urbonas and
Stahre, 1993). Detention ponds are in effect small dams,
and the safety aspects associated with failure and overtop-
ping should also be considered in the design.

In a heavily urbanized landscape there are likely to be
limited opportunities to use the types of detention facilities mentioned below, but use can be made of flat roof
storage, temporary flooding of recreational areas such as
parks and paved precinct areas, and automobile parking
areas. Use of these facilities will obviously cause user
inconvenience and possible hazard, which will need to be

assessed along with the frequency and duration of flooding. Also the users and people responsible for maintenance should be aware of the design function of these detention facilities so that they do not take measures to prevent their flooding.

Extended-Detention Dry Ponds

These temporarily detain a portion of stormwater runoff for up to 48 h (a 24-h limit is more common), using an outlet control. They provide: moderate but variable removal of particulate pollutants; negligible soluble pollutant removal; and quick accumulation of debris and sediment (Metropolitan Washington Council of Governments, 1992b). The performance can be enhanced by use of a forebay to allow sedimentation and easier removal from one area. Many dry ponds that were originally intended for flood control can be modified or retrofitted to serve as wet ponds, thereby providing the additional benefit of removing pollutants as well.

Wet Ponds

These have a permanent pool of water for treating incoming stormwater runoff. Wet ponds have capacity greater than the permanent pond volume, which permits storage of the influent stormwater runoff and controlled release of the mixed influent and permanent pond water. They can provide moderate to high removal of particulate and soluble pollutants and reliable removal rates with pool sizes ranging from 0.5 to 1.0 in. (12.7–25.4 mm) of storm runoff per impervious acre (Metropolitan Washington Council of Governments, 1992b). Wet ponds offer better removals and less maintenance than dry ponds but need to be well designed to ensure they are a benefit to an area and do not cause aesthetic, safety, or mosquito breeding problems. The performance and maintenance requirements can be helped by installing a forebay to trap sediments and allow easier removal, and through use of a fringe wetland on a shallow water bench around the pond perimeter.

There are several variations and combinations that can be used for the above detention systems to enhance the stormwater treatment and/or better suit local conditions. Further details on the design, performance, maintenance, and any special requirements/problems, are available (Metropolitan Washington Council of Governments, 1992b, 1987; Wanielista and Yousef, 1992; and Urbonas and Stahre, 1993).

Infiltration Practices

Infiltration practices have a high potential for controlling stormwater runoff by disposal at a local site level. However, the soil and water table conditions have to be suitable, a sufficiently conservative design has to be used, and adequate maintenance has to be undertaken to minimize the possibility of system failure. The importance of using only suitable sites together with adequate design and maintenance cannot be overstressed for this BMP. Another important aspect is the potential for groundwater pollution. Dissolved pollutants that show little association with solids would be the immediate concern but other pollutants could be more of a problem in the long term. Sandy soils generally have high infiltration rates and a potential to filter the stormwater well. However, they are unlikely to provide good removals through sorption or ion exchange. Soil with a high organic content is likely to offer better capacity to absorb pollutants but at a slower infiltration rate.

Infiltration in its simplest form involves maximizing the pervious area of ground available to allow infiltration of stormwater and minimize the storm runoff. This can be enhanced by directing storm runoff from impervious paved and roof areas to pervious areas, assuming sufficient infiltration capacity exists. Regulations that encourage the incorporation of a high proportion of pervious areas, particularly for new developments, can be effective.

Infiltration Trenches

These are shallow, excavated trenches that have been backfilled with stone to create an underground reservoir. Stormwater runoff that is diverted into the trench gradually exfiltrates from the trench into the surrounding soil and, in many cases, eventually to the water table. There are no real performance data on infiltration trench removals but they are believed to have a good capacity to remove particulate pollutants and a moderate capability to remove soluble pollutants. Variations on this system include the use of perforated pipes to allow exfiltration and conveyance or storage of stormwater in excess of the filtration rate. Clogging of infiltration trenches is the most common cause of their failure. It is important to protect them from sediment loads during and after construction until the surrounding runoff area has developed ground cover to minimize erosion and sediment transport. The system can be enhanced and clogging reduced to providing pretreatment in the form of grass filter strips to filter particulates out of the storm runoff before reaching the infiltration trench (Metropolitan Washington Council of Governments, 1992b).

Infiltration Basins

These are similar to dry ponds (unlined), except that infiltration basins have an emergency spillway only and no standard outlet structure. The incoming stormwater runoff is stored until it gradually exfiltrates through the soil of the basin floor. The comments made above about infiltration trenches will also apply to infiltration basins. Additionally, unlined detention ponds will allow some degree of infiltration.

Porous Pavement

This is a permeable, specially designed, concrete or asphalt mix that provides an alternative to conventional pavement, allowing stormwater to percolate through the porous pavement into a deep gravel storage base area that also acts as a subsurface foundation. The stored storm runoff then gradually exfiltrates into the surrounding soil. In areas where soil has a slow infiltration rate, subsurface piping may be installed to direct the stormwater away. Field studies have shown that porous pavement systems can remove significant levels of both soluble and particulate pollutants (Metropolitan Washington Council of Governments, 1992b; U.S. Environmental Protection Agency, 1981b, 1980). The previous caution about infiltration BMPs affecting the groundwater also apply here. This system tends to be used in areas such as parking areas with gentle slopes and relatively light traffic. This can be an attractive alternative at sites that lack areas to form detention ponds or provide sufficient pervious areas. Sediment loads will clog the surface and should be avoided; this will be particularly important during construction. Also, the regular maintenance of cleaning the surface should be done. On some installations the gravel bed/storage layer has been extended beyond the plan limits of the pavement and returned up at the edge of the pavement. This can enable, with suitable design, any excess storm runoff to be collected by the perimeter gravel. Construction costs of a porous-pavement parking lot will be approximately equal to that of a conventional-pavement parking lot requiring stormwater inlets and subsurface piping (Field et al., 1993).

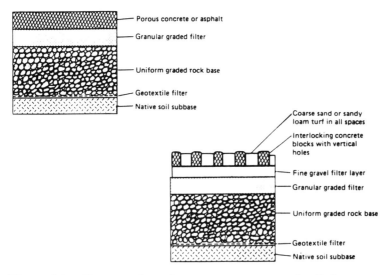

Figure 4.4.   Cross-section of porous pavement and cellular porous pavement (Urbonas and Stahre, 1993).

Another type of porous pavement is constructed using molecular interlocking blocks with open cells that are placed over a deep stone storage base similar to the above porous pavements. This is illustrated in Figure 4.4.

INSTALLED DRAINAGE SYSTEM

The goal of upstream BMPs is to provide sufficient stormwater control to ensure that further downstream treatment is not needed. However, particularly for urban areas it is highly unlikely that this goal will be achieved and further drainage system and end-of-pipe controls will need to be considered.

Control practices that can be applied to the drainage system are relatively limited, especially for existing systems, and involve the following items:

removal of illicit or inappropriate cross-connections

catchbasin cleaning

critical source area treatment devices

infiltration

in-line storage

off-line storage

Many of the control options are similar to those used for CSO control, and in the case of new developments there is the option to install either separate or combined sewer systems. A combined sewer system with treatment is likely to provide the most effective solution in an urban/commercial environment where BMPs are unlikely to provide sufficient reduction of urban storm runoff pollution. In less urbanized areas with strongly enforced BMPs and public support there is a much greater possibility that the downstream stormwater in a separate system will need no further treatment.

For new combined or separate systems, advantage can be taken of increasing the pipe size and gradient to provide in-line storage and self-cleaning, respectively. This will incur an additional cost, which should be relatively small, but the feasibility will be subject to site conditions and available hydraulic head. Existing separate (and combined) drainage systems can be modified for in-line storage by the addition of flow control devices (weirs, flow regulators, etc.).

Established urban areas with separate stormwater drainage systems are most likely to have an existing stormwater pollution problem that needs to be rectified. The following section covers some of the options available.

Illicit or Inappropriate Cross-Connections

This type of control was discussed in the last section, but also is treated here because of its close relation to the drainage system. Identification and removal of illicit or inappropriate connections may provide a partial or complete solution but will be time consuming and costly, with no guarantee of success. Depending on the likely magnitude of the cross-connection problem, it is worth considering the alternative of accepting the pollution problem and providing treatment. If this decision is made early in the investigation there is the potential to maximize the use of resources on the treatment option.

Catchbasin Cleaning

A catchbasin captures settled solids by means of a sump below the outlet invert, and captures floatables by means of a baffle or inverted pipe, and is distinct from an inlet, which has no sump. Pollution control performance is variable, with the trapped liquid generally having a high dissolved pollutant content, which is purged from catchbasins during a storm event, contributing to intensification of the stormwater runoff pollutant loading. Countering this negative impact is the removal of pollutants associated with the settled solids and floatables (e.g., heavy metals and organics) retained in and subsequently cleaned from the basin (U.S. Environmental Protection Agency, 1977b). A regular cleaning schedule is important to maintain the catchbasin performance with a frequency such that sediment buildup is limited to 40–50% of the sump capacity (U.S. Environmental Protection Agency, 1977b), or with cleaning at least twice a year, depending upon conditions.

In a study (U.S. Environmental Protection Agency,

1983a) conducted in West Roxbury, Boston, Massachusetts, three catchbasins were cleaned and four runoff
events were monitored at each catchbasin. The average
pollutant removals per storm are shown in Table 4.3. The
same study also looked at the effectiveness of screening
the stormwater runoff through U.S. standard number-8
brass mesh installed in the three catchbasins. The results
indicated that screens offered a slight gain in overall pollutant removal efficiency for catchbasins. The screens
were effective for the removal of coarse material that could
cause aesthetic problems in the receiving water, but the
potential for clogging and decomposition of trapped material reduced their value unless weekly cleaning was
carried out. The present increased emphasis on stormwater management has resulted in a review of the role that
screening at inlets and catchbasins can play. The city of
Austin, Texas, has developed its own form of inlet filter
(Captur™), which is a relatively coarse screen for removal
of larger stormwater debris. Others (e.g., Emcon North
West) have developed screens utilizing filter material (5–
100 μ) for removal of SS. The Storm and Combined Sewer
Pollution Control Research Program of the U.S. EPA,
through the University of Alabama at Birmingham, is
presently evaluating a number of inlet or catchbasin

Table 4.3. Pollutants Retained in
Catchbasin

Constituent	% Retained
SS*	60–97
Volatile SS	48–97
COD†	10–56
BOD5‡	54–88

*Suspended solids
†Chemical oxygen demand
‡5-day biochemical oxygen demand
(EPA-600/2-83/043)

screening/filtering devices (U.S. Environmental Protection Agency, 1992c).

Critical Source Area Treatment Devices

Research into the source of stormwater pollutants has shown that certain critical source areas can contribute a significant portion of the total urban storm runoff pollutant load (Pitt et al., 1991, 1993). Treatment of the critical source areas can therefore offer the potential for a greater benefit to reduce downstream pollutant loads. Potential critical source areas include: vehicle service, garage, or parking areas; storage and transfer yards; and industrial materials handling areas exposed to precipitation.

Sand Filters
    These use a bed of sand through which the storm runoff is filtered prior to discharge to the drainage system or ground infiltration. Sand filters can offer high removal rates for sediment and trace metals, and moderate removals for nutrients, BOD, and fecal coliform (Metropolitan Washington Council of Governments, 1992). The arrangement of the sand filter bed can vary from an open pit with perforated pipes under the sand bed, as shown in Figure

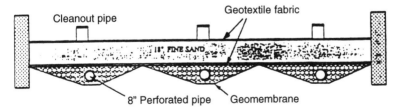

Figure 4.5.   Conceptual design of a sand filter system (MWCG, 1992b).

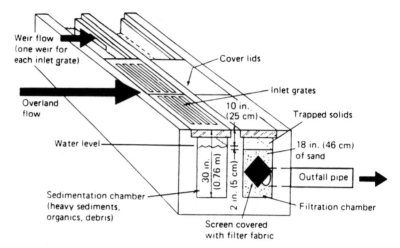

Figure 4.6.   Sand filter stormwater inlet (Urbonas and Stahre, 1993).

4.5, to a more sophisticated trench stormwater inlet, as shown in Figure 4.6, which includes a sediment chamber, weir, and sand filter chamber. Washington, D.C., has installed a few sand filters in chambers in the line of the drainage pipes for treatment of urban storm runoff. The storm runoff passes along the drainage pipe, enters the chamber, passes through the sand filter bed, and returns to the drainage pipe. An overflow bypass is incorporated in the chamber to handle flows in excess of the filter bed capacity.

Maintenance of sand filter beds involves removal of debris from the surface, replacement of the top layer of sand, and raking of the surface. The frequency of this maintenance will be controlled by the rate of accumulation of filtered material.

Oil-Grit Separators

These are usually three-stage underground chambers designed to retain storm runoff, remove heavy particulate by settling, and remove hydrocarbons by trapping of floating material or adsorption onto settled solids. They have limited pollutant removal capability and appear only to trap coarse-grained solids and some hydrocarbons. Removal of silt and clay, nutrients, trace metals, and organic matter is expected to be slight. Without regular clean-out maintenance (e.g., every 3 months), resuspension is likely to limit any long-term removal.

Enhanced Treatment Device

Research is currently being conducted to develop a treatment device for runoff generated by small but critical toxicant source areas. This will consist of a relatively small chamber filled with plastic, hollow slotted media to promote cascading and aeration of the inflow and volatilization of volatile compounds, together with a sump to collect any heavier solids that settle out. The first chamber will then feed the runoff into a second, sedimentation chamber incorporating tube or plate settling for enhancing sedimentation with floating sorption pillows to remove floating oil and grease. This chamber may also be fitted with

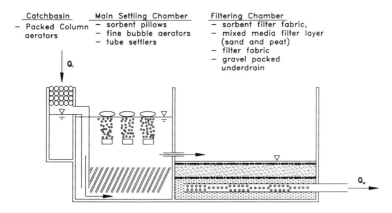

Figure 4.7. Multichambered enhanced treatment device.

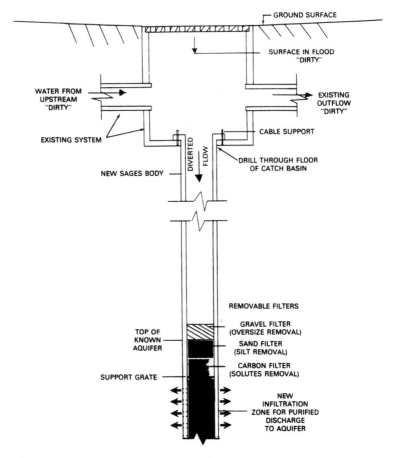

Figure 4.8.   SAGES Unit (931026 Ontario Limited, J. Van Egmond).

aeration facilities depending on the results of the demonstration. The final chamber will contain a sand filter bed, which may also be enhanced with either a homogeneously mixed layer of sand and peat, or a granular activated carbon layer to improve removals (U.S. Environmental Protection Agency, 1992c). This treatment device is shown in Figure 4.7.

The above research study will also involve installation and monitoring of a filtering and pre-infiltration device

(SAGES), shown in Figure 4.8. This device is intended to provide a high level of filtration treatment to the storm runoff prior to local infiltration into the ground.

Infiltration

New installations offer the possibility of using porous conveyance pipes to promote infiltration, but this can be recommended only where the soil and water table conditions are suitable and stormwater pollutants will not cause a problem.

In-line Storage

This is the use of the unused volume in the drainage system network of pipes and channels to store storm runoff. In-line storage capacity can also be provided by storage tanks, basins, tunnels, or surface ponds, which are connected in-line to the conveyance network. To gain maximum benefit from in-line storage it should be combined with some form of treatment; otherwise, only flow attenuation will be achieved. The in-line storage is unlikely to offer any treatment in itself through settling, as the intent will be to make the system self-cleaning to reduce maintenance requirements. However, if the storage is combined with an end-of-pipe treatment, the flow attenuation will help equalize the load to the treatment process, hence optimizing the size of the treatment plant and costs.

The concept of combining storage and treatment to minimize the storage and treatment capacity required and hence optimize the cost to control polluted stormwater is an important relationship. Further cost-effective solutions might be found if existing treatment facilities can be used, such as connection to an existing wastewater system. This is discussed in more detail later in the chapter, as the storage does not necessarily need to be provided by in-line storage.

Even without treatment, flow attenuation will help equalize the loading to be assimilated by the receiving water and reduce the peak flows and consequent erosion in the receiving stream. This can have a major benefit in reducing disturbance of the aquatic ecosystem.

The degree to which the existing conveyance system can be used for storage will be a function of: the pipe sizes that will provide the storage volume; the pipe gradient (relatively flat pipes are likely to provide the most storage capacity without susceptibility to flooding of low areas); suitable locations for installation of control devices such as weirs; and the reliability of the installed control. It will be essential that accurate details of the existing system be collected from field surveys and as-built drawings. This will allow the storage capacity, numbers, and locations of controls, and risk of upstream flooding to be assessed. This will also be invaluable in new drainage system design, where conveyance pipes and channels can be upsized and hydraulic controls can be designed into the system for added system storage and routing.

Controls used to restrict flow causing a backup and storage in the system fall into two categories, either fixed or adjustable. Fixed systems are likely to be cheaper and require less maintenance but do not offer the flexibility and potential to maximize the storage potential. Adjustable systems can offer the advantage of being connected to a real-time control (RTC) system, which, via a system of rainfall measurements and forecasts, monitoring of stormwater levels in critical sections of the drainage system, and input of this data into a computer system, can be adjusted to hold back or release stormwater to maximize storage capacity of the whole drainage system. RTC systems have been installed and are being further developed to control complex sewerage systems in the CSO field. The sophistication offered by an RTC system is unlikely to offer a cost-effective solution for a separate storm drainage system unless there is a large in-line storage capacity and the stored runoff is to be treated.

Typical examples of fixed and adjustable flow regulators are listed below:

Fixed Regulators	Adjustable Regulators
Orifices	Inflatable dams
Weirs (lateral and longitudinal)	Tilting plate regulators
Steinscrew	Reverse-tainter gates
Hydrobrake	Float-controlled gates
Wirbeldrossel	Motor-operated or hydraulic gates
swirl	
stilling-pond weir	

Some of the above are relatively inexpensive, quick to install, and an effective means of increasing storage. Several publications (Urbonas and Stahre, 1993; U.S. Environmental Protection Agency, 1977a, 1970a, 1970b) on CSO control can provide more information on the above regulators. However, as stated earlier, without treatment the advantage of storage is only in flow attenuation. It should also be noted that some of the above regulators will concentrate the heavier solids in the stored storm runoff for a more concentrated later release.

Off-line Storage

This refers to storage that is not in-line to the drainage conveyance system. Storage is achieved by diverting flow from the drainage system when a certain flow rate is exceeded. The diverted water is stored until sufficient capacity is available downstream. The off-line storage can be provided by any arrangement of basins, tanks, tunnels, etc., and, if gravity filling and emptying are not possible, will involve pumping the water into or out of storage.

Off-line storage, as with in-line storage, can be designed to be relatively self-cleaning or have facilities to resuspend the settleable solids. Examples can be found in books on stormwater (Metcalf and Eddy, 1981; Field, 1990; Urbonas and Stahre, 1993). Off-line storage can also be used to provide treatment by sedimentation, with the sludge either collected or diverted to a wastewater treatment plant.

Many of the regulators listed under *in-line storage* can be used to divert the flow once the predetermined flow rate has been exceeded. In addition to the above-listed regulators, vortex and helical bend regulators/concentrators can be used. As their name suggests, they will concentrate the heavier solids into the underflow, which will continue to be conveyed along the drainage pipes. Therefore end-of-pipe treatment is required if this concentrated pollutant load is to be prevented from reaching the receiving water. The regulator/concentrator can offer advantages for end-of-pipe treatment when the flow needs to be regulated to prevent the treatment capacity from being exceeded. End-of-pipe treatment can be satellite or central treatment. This is discussed further in the *end-of-pipe treatment* section.

Flow Balance Method (FBM)

The system provides a means of storing discharged urban storm runoff in the receiving water. This allows either pumpback for treatment, when capacity is available, or treatment of the runoff by sedimentation until the next storm runoff event displaces the stored volume. The method was first developed in Sweden (Soderlund, 1988) as a means of protecting lakes against pollution from stormwater runoff and has since been demonstrated for control of CSO in a marine receiving water in Jamaica Bay, New York City, New York (Field et al., 1990, Forndran et al., 1991).

Storage in the receiving water is achieved by forming a tank using flexible plastic curtains suspended from pontoons. The curtains are anchored to the receiving water bottom by concrete weights and the base of the tank is formed by the receiving water bed. The relatively low cost of the materials and construction gives this system cost advantages over conventional concrete and steel tank systems (estimated to be one-fifth to one-tenth the cost), requires only a minimal amount of land space for controls and access, and has flexibility to expand the volume if required at a later date.

The Swedish freshwater-lake installations use a connected system of bays with openings between adjacent/sequential tanks to facilitate movement of the stormwater and lake water between tanks. Lake water can enter and leave these FBMs via the last tank in the series, which has an opening to the lake. Plug flow set up by the

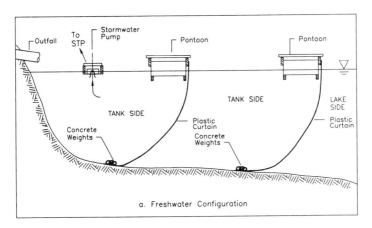

Figure 4.9a. Flow balance method (FBM)—freshwater configuration.

discharging stormwater displaces the lake water from the first to the second bay and on down the line until the discharge finishes or each bay is filled with stormwater (i.e., stormwater has to pass through all of the bays to gain access to the lake). A reverse flow sequence occurs during pumpback of the stormwater to the wastewater treatment plant (WWTP). Figure 4.9a shows the FBM freshwater system.

Sweden has invested in three of these installations, which have all been in operation for a number of years. The systems have withstood wave action up to 3 ft (0.9 m) as well as severe icing conditions. If a wall is punctured, patching is easily accomplished and general maintenance has been found to be inexpensive. The FBM has been successfully demonstrated in these lakes, resulting in improved water quality in the lakes (Soderlund, 1988; Pitt and Dunkers, 1993).

The marine FBM demonstration (see Figure 4.9b) utilizes a different operating principle, of density difference for displacement instead of plug flow. One tank is used

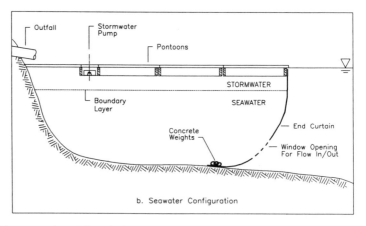

Figure 4.9b. Flow balance method (FBM)—seawater configuration.

and the seawater is displaced vertically by the lower-density CSO influent floating on the higher-density sea water, and hence forming a stratified layer of CSO above the lower seawater layer. The demonstration project is in two phases. The first phase concentrated on proving the feasibility of the system concept to displace seawater, form a stable CSO layer, pump the CSO back to the WWTP, and allow for the system structure to withstand a marine environment including tidal exchange, freezing, and coastal storms. The second phase (currently in progress) expands the system capacity from 0.41 Mgal (1550 m^3) to 2 Mgal (7570 m^3), and concentrate on monitoring the system performance (U.S. Environmental Protection Agency, 1990).

During the two-year demonstration project, the system withstood the marine environment (the FBM was located in a relatively sheltered seawater creek), with no structural damage or material degradation observed. The system was exposed to tidal ranges up to 7 ft (2 m), winds gusting to 40 mph (64 km/h), and icing conditions. The system was shown to retain CSO in a stratified layer that remained relatively stable and could be pumped back to the WWTP. The FBM proved effective in trapping floatable material and a means of floatable material removal is part of the next phase. Pumpback of the settled solids from the FBM bed had been incorporated into both phases.

It is important to note that although an FBM can offer a cost-effective and quick-to-construct storage facility it requires a suitable location and does have limits on its performance. There will be a certain amount of mixing with the receiving water, not all of the stored volume will be pumped back, and any settleable solids will settle out of the stored storm runoff (regular pumpback of the accumulated sediment would help overcome this problem). The low cost and quick construction potential of the FBM could favor the use of this system as a temporary measure in cases of a severe problem that needs attention. The FBM does use the existing natural receiving water and therefore requires all the necessary permits involved in such situations.

Maintenance

In order for the drainage system and the controls to work efficiently they should all be regularly maintained. This will generally consist of removing sediments from control devices, flushing drainage lines, and carrying out general inspections to identify any problems. Regular maintenance will also minimize any buildup of material that could be flushed out by a surge from a large storm event and thereby minimize the shock loading caused by intermittent storm events.

END-OF-PIPE TREATMENT

Biological Treatment

Biological treatment provides a means of removing organic pollutants from the storm runoff either aerobically or anaerobically. For this treatment to be effective, the systems must be operated continuously to maintain an active biomass or be able to borrow the biomass from a system that does operate continuously. Biological processes are relatively sensitive and can be affected by the variable flow conditions and the relatively high concentration of nonbiodegradable solids in storm runoff. These factors tend to make high-rate physical treatment processes more suitable for stormwater applications, with their ability to handle high and variable flow rates and solids concentrations.

An exception to the above is the rotating biological

contactor (RBC), which is less susceptible to overloading shock loads compared to other biological systems, such as trickling filters and waste activated sludge processes. RBCs have achieved high removals at flows 8 to 10 times their base flow for CSO treatment (U.S. Environmental Protection Agency, 1974d). However, RBCs, like all biological processes, need a food source to keep the microbes alive during extended dry periods and therefore have their limitations. The remainder of this section will therefore concentrate on the physical/chemical treatment processes that tend to be more suitable for treatment of stormwater.

Use of Existing Treatment Facilities

As stated earlier any use of existing facilities is likely to provide cost-effective treatment, provided an economic means of connecting the stormwater drainage system to the facility is possible. Use of spare capacity at wastewater treatment plants is one option, particularly if storage can be provided to equalize the storm runoff load. Even if the biological system has very little capacity the primary treatment systems can often function well at much higher overflow rates, which, if combined with disinfection of the discharged storm runoff, will offer significant treatment. Stormwater also tends to have a higher percentage of heavier solids than sanitary sewage, which will benefit removals at higher overflow rates.

An alternative could be to construct additional primary treatment at a WWTP to run in series with existing facilities during DWF, for improved treatment of DWF, and run in parallel during wet-weather flow. Use of any storage facilities, either at an end-of-pipe or an upstream location, could provide treatment by sedimentation or storage to be released when treatment capacity is available.

Physical/Chemical Treatment

These processes generally offer good resistance to shock loads, ability to consistently produce a low suspended solids (SS) effluent, and adaptability to automatic operation. They are generally, with the exception of high-gradient magnetic separation and powdered activated carbon, suitable only for removal of SS and associated pollutants. The extent of removals will depend on the SS characteristics and the level of treatment applied. The physical/chemical systems to be discussed are:

screening

filtration

dissolved air flotation

high-gradient magnetic separation

powdered activated carbon–alum coagulation

disinfection

swirl concentrators/regulators

Screening

Screens can be divided into four categories, with the size of the SS removed directly related to the screen aperture size:

Screen Type	Opening Size
Bar screen	>1 in. (>25.4 mm)
Coarse screen	3/16–1 in. (4.8–25.4 mm)
Fine screen	1/250–3/16 in. (0.1–4.8 mm)
Microscreen	<1/250 in. (<0.1 mm)

Bar and coarse screens have been used extensively in WWTP at the headworks to remove large objects. Depending on the level of treatment required for the storm runoff the smaller-aperture coarse screens may be sufficient; however, a higher level of treatment can be achieved using the bar and coarse screens in conjunction with the fine or microscreens. Design of screens can be similar to that for WWTP and CSO, but with consideration for stormwater characteristics of intermittent operation and possible very high initial loads that may not reflect WWTP operation characteristics. A self-cleaning system should be included for static screens to save manual cleaning during storm events together with automatic startup and shutdown. Catenary screens fall into the coarse-screen category; they are rugged and reliable and commonly used for CSO facilities. Therefore they are likely to be a good screen for use with storm runoff.

Table 4.4 lists devices that fall into the fine screen and microscreen category, and were developed and used for SS removal from CSO. With no information on screening of separate stormwater, the information on screening CSO is a good starting point and the information given below is from CSO studies. Design parameters for static screens, microstrainers, drum screens, disc screens, and rotary screens are presented in Tables 4.5, 4.6, and 4.7. The removal efficiency of screening devices is adjustable by changing the aperture (size of opening) of the screen placed on the unit, making these devices very versatile. In other words, the efficiencies of a screen treating a waste with a typical distribution of particle sizes will increase as the screen aperture decreases.

Solids removal efficiencies are affected by two mechanisms: straining by the screen, and filtering of smaller particles by the mat deposited by the initial straining. Suspended matter removal will increase with increasing thickness of filter mat because of the filtering action of the mat itself, which is especially true for microstrainers. This will also increase the headloss across the screen. A study

Table 4.4. Description of Screening Devices Used in CSO Treatment

Type of Screen	General Description	Process Application	Comments
Drum screen	Horizontally mounted cylinder with screen fabric aperture in the range of 100–843 microns. Operates at 2–7 r/min.	Pretreatment	Solids are trapped on inside of drum and are backwashed to a collection trough.
Microstrainers*	Horizontally mounted cylinder with screen fabric aperture 23–100 microns. Operates at 2–7 r/min.	Main treatment	Solids are trapped on inside of drum and are backwashed to a collection trough.
Rotostrainer	Horizontally mounted cylinder made of parallel bars perpendicular to axis of drum. Slot spacing in the range of 250–2500 microns. Operates at 1–10 r/min.	Pretreatment	Solids are retained on surface of drum and are removed by a scraper blade.

Disc strainer	Series of horizontally mounted woven wire discs mounted on a center shaft. Screen aperture in the range of 45–500 microns. Operates at 5–15 r/min.	Pretreatment, main treatment, or posttreatment of concentrated effluents.	Unit achieves a 12–15% solids cake.
Rotary screen	Vertically aligned drum with screen fabric aperture in the range of 74–167 microns. Operates at 30–65 r/min.	Main treatment	Splits flow into two distinct streams: unit effluent and concentrate flow, in the proportion of approximately 85:15.
Static screen	Stationary inclined screening surface with slot spacing in the range of 250–1600 microns.	Pretreatment	No moving parts. Used for removal of large suspended and settleable solids.

*A vertically mounted microstrainer is available, which operates totally submerged at approximately 65 r/min. Aperture range is 10–70 microns. Solids are removed from the screen by a sonic cleaning device.
(EPA-600/8-77/014)

Table 4.5. Design Parameters for Static Screens

Hydraulic loading, gal/min/ft of width	100–180
Incline of screens, degrees from vertical	35*
Slot space, microns	250–1600
Automatic controls	None

*Bauer Hydrasieves™ have 3-stage slopes on each screen: 25°, 35°, 45°.
gal/min/ft × 0.207 = 1/m/s
(EPA-600/8-77/014)

in Philadelphia (Field, 1972) showed (on a 23-µ-aperture microscreen [Microstrainer]) that with a large variation in the influent SS, the effluent SS stayed relatively constant (e.g., if a 1000-mg/L influent SS gave a 10-mg/L effluent SS, then a 20-mg/L influent SS would still give a 10-mg/L effluent SS). Accordingly, treatment efficiencies vary with influent concentration.

Microscreens and fine screens remove 25–90% of the SS, and 10–70% of the BOD_5, depending on the screen aperture used and the wastewater being treated. The above Philadelphia study showed that improved removals and increased flux densities (hydraulic loadings) are possible using polyelectrolyte addition. This is also likely to be the case with storm runoff, but laboratory coagulation studies would be needed to find the best polyelectrolyte and dosage for the particular storm runoff characteristics. The optimum dosage will change with changes in the storm runoff characteristics, requiring some form of automated monitoring (e.g., SS monitoring) for adjustment of dosage or setting of an average effective dosage.

More detailed descriptions of the various screening devices are available in the literature (U.S. Environmental Protection Agency, 1977a; Metcalf and Eddy, 1981; Field, 1990; Water Environment Federation, 1992).

Filtration

Dual-media high-rate filtration (DMHRF) (>8 gpm/ft^2 [20 m^3/m^2/h]) removes small particulates that remain

Table 4.6. Design Parameters for Microstrainers, Drum Screens, and Disc Screens

Parameter	Microstrainers	Drum Screen	Disc Screen
Screen aperture, microns	23–100	100–420	45–500
Screen material	stainless steel or plastic	stainless steel or plastic	wire cloth
Drum speed, r/min			
Speed range	2–7	2–7	5–15
Recommended speed	5	5	—
Submergence of drum, %	60–80	60–70	50
Flux density, gal/ft²/min of submergence screen	10–45	20–50	20–25
Headloss, in.	10–24	6–24	18–24
Backwash			
Volume, % of inflow	0.5–3	0.5–3	—*
Pressure, lb/in.²	30–50	30–50	—

*Unit's waste product is a solids cake of 12–15% solids content
$gal/min/ft^2 \times 2.445 = m^3/h/m^2$
$in. \times 2.54 = cm$
$ft \times 0.305 = cm$
$lb/in.^2 \times 0.0703 = kg/cm^2$
(EPA-600/8-77/014)

Table 4.7. Design Parameters for Rotary Screens

Screen aperture, microns	
Range	74–167
Recommended aperture	105
Screen material	stainless steel or plastc
Peripheral speed of screen, ft/s	14–16
Drum speed, r/min	
Range	30–65
Recommended speed	55
Flux density, gal/ft^2/min	70–150
Hydraulic efficiency, % of inflow	75–90
Backwash	
Volume, % of inflow	0.02–2.5
Pressure, lb/in.^{2}	50

ft/s × 0.305 = m/s
gal/ft^2/min × 2.445 = m^3/m^2/h
lb/in^2 × 0.0703 = kg/cm^2
(EPA-600/8-77/014)

after screening and floc remaining after polyelectrolyte and/or coagulant addition. As implied, this provides a high level of treatment that can be applied after screening, together with automated operation and limited space requirements. To be most effective, filtration through media that are graded from coarse to fine in the direction of flow is desirable. A single filter material with constant specific gravity cannot conform to this principle because backwashing of the bed automatically grades the bed from coarse to fine in the direction of washing; however, the concept can be approached by using a two-layer bed. A typical case is the use of coarse anthracite particles on top of less-coarse sand. Because anthracite is less dense than sand, it can be coarse and still remain on top of the bed after the backwash operation. Typically a unit is comprised of 5 ft of No. 3 anthracite (effective size 0.16 in. [4.0 mm]) placed over 3 ft of No. 612 sand (effective size 0.08 in. [2.0 mm]). This arrangement was shown to be superior to both coarser and finer media tested separately (U.S. Environmental Protection Agency, 1972b). Another alternative would be an upflow filter, but these units are limited in that they cannot accept high hydraulic loadings (filtration rates).

The principal parameters to be evaluated in selecting a DMHRF system are the media size, media depth, and filtration rate. Because much of the removal of solids from the water takes place within the filter media, their structure and composition are of major importance. Too fine a medium may produce a high-quality effluent, but also may cause excessive headlosses and extremely short filter runs. On the other hand, media that are too coarse may fail to produce the desired clarity of the effluent. Therefore, the selection of media for DMHRF should be made by pilot testing using various materials in different proportions and at different flow rates. Depth of media is limited by headloss and backwash considerations. The deeper the bed, the greater is the headloss and the harder the cleaning. However, there should be sufficient bed depth to retain the removed solids without breakthrough during the filter run period at the design hydraulic loading.

Information is available on the use and design of DMHRF for treatment of drinking water, but a number of pilot studies have also been done, using CSO, that should provide more-relevant information. The studies (U.S. Environmental Protection Agency, 1972b, 1979a, 1979b) used 6-, 12-, and 30-in. (15, 30, and 76 cm) diameter filter columns, with anthracite and sand media with and without various dosages of coagulants and/or polyelectrolytes. A preliminary (420 μm) screening process was used upstream of the DMHRF to extend the treatment run time before backwashing. It was found that SS removal increased as influent SS concentration increased, and decreased as hydraulic loading increased.

Removal efficiency for the filter unit was about 65% for SS, 40% for BOD_5, and 60% for chemical oxygen demand (COD). The addition of polyelectrolyte increased the SS removal to 94%, the BOD5 removal to 65%, and the COD removal to 65%. The length of filtration run averaged 6 h at a hydraulic loading of 24 gpm/ft^2 (59 m^3/m^2/h).

Tables 4.8, 4.9, and 4.10 show removals of SS, BOD5, and heavy metals for a study in New York, New York (U.S. Environmental Protection Agency, 1979a). Design parameters for DMHRF are presented in Table 4.11 (U.S. Environmental Protection Agency, 1977a).

Dissolved Air Flotation (DAF)

This is a unit operation used to separate solid particles or liquid droplets from a liquid phase. Separation is brought about by introducing fine air bubbles into the liquid phase. As the bubbles attach to the solid particles, the buoyant force of the combined particles and air bubbles is great enough to cause the particles to rise. Once the particles have floated to the surface, they are removed by skimming. The most common process for forming the air bubbles is to dissolve air into the waste stream under pressure and then release the pressure to allow the air to come out of solution. The pressurized flow carrying the dissolved air to the flotation tank is either: (1) the entire stormwater flow, (2) a portion of the stormwater flow (split flow pressurization), or (3) recycled DAF effluent.

Higher overflow rates (1.3–10.0 gpm/ft^2 [3.2–25 m^3/m^2/h]) and shorter detention times (0.2–1.0 h) can be used for DAF when compared to conventional settling (0.2–0.7 gpm/ft^2 [0.5–1.7 m^3/m^2/h]; 1.0–3.0 h). Studies for CSO have shown that a treatment system consisting of screening (using a 297-μ aperture with a hydraulic loading rate of 50 gpm/ft^2 (122.3 m^3/m^2/h)) followed by DAF can offer an effective level of treatment (U.S. Environmental Protection Agency 1977c, 1979c). The basis of the system being that the screening removes the particles that are too heavy for the air bubbles to carry, and the DAF system removes the floating, neutral-buoyancy, and remaining

Table 4.8. CSO–DMHRF Average SS Removals (New York, NY)

	Plant Influent (mg/l)	Filter Influent (mg/l)	Filter Effluent (mg/l)	Filter Removals (%)	System Removals (%)
No chemicals	175	150	67	55	62
Poly only	209	183	68	63	67
Poly and alum	152	142	47	67	69

(EPA-600/2-79/015)

Table 4.9. CSO–DMHRF Average BOD5 Removals (New York, NY)

	Plant Influent (mg/l)	Filter Influent (mg/l)	Filter Effluent (mg/l)	Filter Removals (%)	System Removals (%)
No chemicals	164	131	96	27	41
Poly only	143	129	84	35	41
Poly and alum	92	85	53	38	43

(EPA-600/2-79/015)

Table 4.10. Removal of Heavy Metals by DMHRF (New York, NY)

	Cadmium	Chromium	Copper	Mercury	Nickel	Lead	Zinc
Average removal, %*	56	50	39	0	13	65	48

*concentration basis
(EPA-600/2-79/015)

Table 4.11. Design Parameters for DMHRF

Filter media depth (ft)		Headloss (ft)	5–30
No. 3 anthracite	4–5		
No. 612 sand	2–3	Backwash	
		Volume (% of flow)	4–10
Effective size (mm)		Air	
Anthracite	4	Rate (standard (ft^3/min/ft^2)	10
Sand	2	Time (min)	10
		Water	
		Rate (gal/ft^2/min)	60
		Time (min)	15–20
Flux density			
(gal/ft^2/min)			
Range	8–40		
Design	24		

ft × 0.305 = m
gal/ft^2/min × 2.445 = m^3/m^2/h
standard ft^3/min/ft^2 × 0.305 = m^3/m^2/min

negative-buoyancy particles. The addition of chemical flocculent in the form of ferric chloride and cationic polyelectrolyte was shown in the above two references to improve the removals. Table 4.12 shows the screening–DAF system design parameters (U.S. Environmental Protection Agency, 1977a).

As with the other treatment processes discussed, there are no data available for treatment of separate storm runoff; however, from the CSO data it would appear that except for sedimentation, screening–DAF is likely to be the most expensive treatment system.

Table 4.12. Screening and DAF Design Parameters

Overflow rate (gal/ft^2/min)	
Low rate	1.3–4.0
High rate	4.0–10.0
Horizontal velocity (ft/min)	1.3–3.8
Detention time (min)	
Flotation cell range	10–60
Floatation cell average	25
Saturation tank	1–3
Mixing chamber	1
Pressurized flow (% of total flow)	
Split-flow pressurization	20–30
Effluent recycle pressurization	25–45
Air-to-pressurized-flow ratio (standard ft^3/min/100 gal)	1.0
Air-to-solids ratio	0.05–0.35
Pressure in saturation tank (lb/in.^{2})	40–70
Float	
Volume (% of total volume)	0.75–1.4
Solids concentration (% dry weight basis)	1–2

gal/ft^2/min × 2.445 = m^3/m^2/h
ft/min × 0.00508 = m/s
standard ft^3/min/100 gal × 0.00747 = m^3/min/100 l
lb/in.^{2} × 0.0703 = kg/m^2
(EPA-600/8-77/014)

High-Gradient Magnetic Separation (HGMS)

This is a relatively new treatment technology for treatment of storm runoff or CSO but has been used successfully for a number of years in the treatment of water for or from industrial processes. A high degree of treatment is possible with this process, which will probably be greater than required to meet permitting requirements alone.

In its simplest form, the high-gradient magnetic separator consists of a canister packed with a fibrous ferromagnetic material that is magnetized by a strong external magnetic field (coils surround the canister). The water to be treated is passed through the canister and the fibrous ferromagnetic matrix causes only a small hydraulic resistance because it occupies less than 5% of the canister volume. Upstream of the canister the water is prepared by binding finely divided magnetic seed particles, such as magnetic iron oxide (magnetite), to the nonmagnetic contaminants. Binding the magnetic seed is accomplished in two general ways: adsorption of the contaminant to the magnetic seed and chemical coagulation (alum).

The magnetic particles are trapped on the edges of the magnetized fibers in the canister as the water passes through. When the matrix has become loaded with magnetic particles, they are easily washed off by turning off the magnetic field and backflushing. Particles ranging in size from soluble through settleable (>0.001 μ) may be removed with this process; design parameters for HGMS are presented in Table 4.13. HGMS can offer rapid filtration for many pollutants with greater efficiency than for sedimentation because the magnetic forces on the fine particles may be many times greater than gravitational forces. *Urban Stormwater Management and Technology: Update and User's Guide* (U.S. Environmental Protection Agency, 1977a) provides details of bench and pilot scale studies that have been conducted using HGMS to treat CSO (U.S. Environmental Protection Agency, 1977a).

Table 4.13. Preliminary Design Parameters for
High-Gradient Magnetic Separators

Magnetic field strength, kG*	0.5–1.5
Maximum flux density, gal/ft²/min	100
Maximum detention time, min	3
Matrix loading, g solids/g of matrix fiber	0.1–0.5
Magnetite addition, mg/l	100–500
Magnetite to SS ratio	0.4–3.0
Alum reduction, mg/l	
Range	90–120
Average	100
Polyelectrolyte addition, mg/l	0.5–1.0

*Kilogauss
gal/ft²/min × 2.445 = m³/m²/h
(EPA-600/8-77/014)

For HGMS and all other treatments that involve an additive to enhance the solids removal, there is a need to accommodate the variation in storm runoff SS concentrations. This will require automatic monitoring and adjustment of the additive dosage for efficient operation.

Powdered Activated Carbon–Alum Coagulation

A treatment option that has the potential to remove dissolved organics in the use of powdered activated carbon with alum added to aid in subsequent clarification. This was demonstrated at a 100,000 gpd (379 m³/d) pilot unit in Albany, New York (Field, 1990; U.S. Environmental Protection Agency, 1973c), using municipal sewage and CSO. A short flocculation period followed the addition of alum with settling of solids by gravity and disinfection of the effluent or filtering (tri-media) and disinfection prior to discharge.

Carbon regeneration in a fluidized bed furnace and alum recovery from the calcined sludge were also demonstrated, as was reuse of the reclaimed chemicals. Average carbon losses per regeneration cycle were 9.7%. Average removals were in excess of 94% for COD, 94% for BOD_5, and 99% for SS with no filtration.

Disinfection

Disinfection is generally practiced at WWTPs to control pathogenic microorganisms. The development of disinfection techniques and measurement of their effectiveness to kill pathogens has been mainly derived from the sanitary wastewater field, where the concern has been to measure the presence of fecal contamination and ability to kill any pathogens and viruses of human origin. Because it is both difficult and expensive to isolate and measure specific pathogens in water, methods were developed to monitor certain indicator organisms—i.e., microorganisms indicative of the presence of fecal contamination. Bacteria of the total coliform (TC) group became the generally accepted indicator for fecal pollution, but they include different genera that do not all originate from fecal wastes (e.g., *Citrobacter*, *Klebsiella*, and *Enterobacter*).

An improvement over the TC test is the more selective fecal coliform (FC) test, which selects primarily for *Klebsiella* and *Escherichia coli (E. coli)* bacteria. *E. coli* is the bacterium of interest because it is a consistent inhabitant of the intestinal tract of humans and other warm-blooded animals. However, the FC test is still not specific to enteric bacteria, and human-enteric bacteria in particular. In 1986 a U.S. EPA publication (U.S. Environmental Protection Agency, 1986) recommended that states "begin the transition process to the new (*E. coli* and *enterococci*) indicators." However, many states still retain the TC and FC criteria, and the most widely used bacteriological criterion in the United States is the maximum of 200 FC/100 mL.

For discharges of separate storm runoff the above criterion is unlikely to give a true indication of the potential risk of infection, as many of these indicator bacteria also originate from soils, vegetation, and animal feces. Stormwater runoff can contain high densities of the non-human indicator bacteria, and epidemiological studies of recreational waters receiving stormwater runoff have found little correlation between indicator densities and swimming-related illnesses (U.S. Environmental Protection Agency, 1983b, 1984a; Calderon et al., 1991). In addition, a number of non-enteric pathogens found in stormwater runoff have been linked to respiratory illnesses and skin infections, a risk that is not assessed by the present fecal indicators.

Although the present standards and indicators are unlikely to reflect the actual human disease contraction potential—i.e., pathogenicity of a storm flow and its receiving water—they are the only practical standards available. Also, urban storm runoff has a high potential to be contaminated by sanitary cross-connections, which would make the standards more relevant. Therefore, until other, more relevant indicators are developed and proven, the present standards should be used but with the caution that they may over- or underestimate the true risk. The paper entitled "The Detection and Disinfection of Pathogens in Storm-Generated Flows" (O'Shea and Field, 1992) covers this subject in more detail.

Conventional municipal sewage disinfection generally involves the use of chlorine gas or sodium hypochlorite as the disinfectant. To be effective for disinfection purposes, a contact time of not less than 15 min at peak flow rate and a chlorine residual of 0.2–2.0 mg/L are commonly recommended. However, a different approach is required for storm runoff, because the flows have characteristics of intermittency, high flow rate, high SS content, wide temperature variation, and variable bacterial quality. Further aspects of disinfection practices that require consideration for storm runoff include the following.

1. A residual disinfecting capability may not be permitted, because chlorine residuals and compounds discharged to natural waters may be harmful to human and aquatic life (e.g., formation of carcinogens, such as trihalomethanes).

2. The coliform count is increased by surface runoff in quantities unrelated to pathogenic organism concentration. Total coliform or fecal coliform levels may not be the most useful indication of disinfection requirements and efficiencies.

3. Discharge points requiring disinfection are often at outlying points on the drainage system and require unmanned, automated installations.

The disinfectant used to treat storm runoff should be adaptable to intermittent use, effective, and safe and easy to dose the effluent with. Table 4.14 shows disinfectants that might be used for storm flow disinfection. Chlorine and hypochlorite will react with ammonia to form chloramines and with phenols to form chorophenols. These are toxic to aquatic life and the latter compounds also produce taste and odor in the water. Chlorine dioxide (ClO_2) does not react with ammonia and completely oxidizes phenols.

Ozone has a more rapid disinfecting rate than chlorine, is effective in oxidizing phenols, and has the further advantage of supplying additional oxygen to the effluent. The increased disinfecting rate of ozone requires shorter contact times and results in a lower capital cost for a contactor, as compared to that for a chlorine contact tank. Ozone does not produce chlorinated hydrocarbons or a long-lasting residual, as chlorine does, but it is unstable and must be generated on-site just prior to application. Therefore, capital investment in a generating plant is required along with the operation and maintenance.

Another disinfection technique that promises short de-

Table 4.14. Characteristics of Principal Storm Flow Disinfection Agents

Characteristics	Chlorine	Hypochlorite	Chlorine Dioxide	Ozone
Stability	Stable	6-month half life	Unstable	Unstable
Reacts with ammonia to form chloramines	Yes	Yes	No	No
Destroys phenols	At high concentrations	At high concentrations	Yes	Yes
Produces a residual	Yes	Yes	Short-lived*	No
Affected by pH	More effective at pH <7.5	More effective at pH <7.5	Slightly	No
Hazards	Toxic	Slight	Toxic; explosive	Toxic

*Chlorine dioxide dissociates rapidly.
(EPA-600/8-77/014)

tention times and the absence of toxic reaction products is the use of ultraviolet (UV) light irradiation. The effectiveness of the early systems was limited for water with high concentrations of solids that tended to attenuate the UV energy. Later systems emit higher-intensity radiation for more-effective treatment. More recently, modulated UV light has been reported to reduce viable bacteria by approximately 100-fold compared to populations observed after similar exposure to UV light that lacked modulation (Bank et al., 1990).

The characteristics of storm runoff (i.e., intermittent and often high flows), together with the need to minimize capital costs for a treatment operation, lend themselves favorably to use of high-rate disinfection. This refers to achieving either a given percent or a given bacterial count reduction through the use of: decreased disinfectant contact time; increased mixing intensity; increased disinfectant concentration; chemicals having higher oxidizing rates; or various combinations of these. Where contact times are less than 10 min, (usually in the range 1–5 min), adequate mixing is a critical parameter, providing complete dispersion of the disinfectant and forcing disinfectant contact with the maximum number of microorganisms. The more physical collisions high-intensity mixing causes, the lower the contact time requirements. Mixing can be accomplished by mechanical flash mixers at the point of disinfectant addition and at intermittent points, or by specially designed plug flow contact chambers containing closely spaced, corrugated parallel baffles that create a meandering path for the wastewater (U.S. Environmental Protection Agency, 1973b).

High-rate disinfection was shown (for CSO) to be enhanced beyond the expected additive effect by sequential addition of Cl_2 followed by ClO_2 at intervals of 15–30 s (U.S. Environmental Protection Agency, 1975a, 1976b). A minimum effective combination of 8 mg/L of Cl_2 followed by 2 mg/L of ClO$_2$ was effective in reducing TC, FC, fecal streptococci, and viruses to acceptable target levels and compared to 25 mg/L Cl$_2$ or 12 mg/L ClO$_2$. It was surmised that the presence of free Cl$_2$ in solution with chlorite ions (ClO$^-_2$ [the reduced state of ClO$_2$]) may cause the oxidation of ClO$^-_2$ back to its original state. This process would prolong the existence of ClO$_2$, the more potent disinfectant.

An equation and concept to enable the effect of high-rate mixing to be taken into account in the disinfection process are provided in U.S. Environmental Protection Agency, 1973b. A velocity gradient (G), as defined in the equation below, is used as a measure of the mixing intensity, but is also a measure of the opportunities for microorganism and disinfectant matter collisions per unit time per unit volume.

$$G = \sqrt{\left(\frac{Energy\ dissipation/Volume}{Viscosity}\right)}$$

The product of velocity gradient and contact time (GT) is the number of opportunities for collisions per unit volume during the contact time.

It is important to note that if high-rate mixing is to be relied upon to provide effective disinfection, the velocity gradient should not reduce if the flow rate reduces—i.e., if the mixing intensity depends on the velocity of flow and not mechanical mixing, then the level of disinfection will be reduced at low flow rates. There will be some offset of this due to longer detention times at lower flow rates but the intensity of mixing will be the more significant parameter. Use of a Sutro weir for the influent and effluent will help maintain the peak-rate velocity at all flow rates.

Swirl Regulators/Concentrators

These are compact flow throttling, and solids separation devices that also collect floatable material. The performance of the swirl device is very dependent on the settling characteristics of the solids in the stormwater. The EPA swirl is most effective at removing solids with characteristics similar to those of grit ($\geq$0.008 in. [0.2 mm] effective diameter, 2.65 specific gravity). It is important to appreciate this aspect of swirl devices and not expect significant removals of fine and low-specific-gravity solids.

The three most common configurations are the EPA swirl concentrator, the Fluidsep™ vortex separator, and the Storm King™ hydrodynamic separator. Although each of the separators is configured differently, the operation of each unit and the mechanism for solids separation are similar. The flow enters the unit tangentially and follows the perimeter wall of the cylindrical shell, creating a swirling, quiescent vortex flow pattern. The swirling action throttles the influent flow, and causes solids to be concentrated at the bottom of the unit. The throttled underflow containing the concentrated solids passes out through an outlet in the bottom of the unit, while the clarified supernatant passes out through the top of the unit. Various baffle arrangements are provided to capture floatables in the supernatant that are then usually carried out in the underflow as the storm subsides and the water level in the swirl unit falls. During low-flow conditions all of the flow passes out via the bottom outlet and only when the flow increases does the throttling effect and buildup of water in the swirl occur.

The solids separation is helped by the flow patterns, with the influent being deflected into a slower-moving inner swirl pattern after one revolution around the perimeter of the swirl unit. Gravity separation occurs as particles follow a "long path" through the outer and inner swirl. Solids separation is also assisted by the shear forces set up between the inner and outer swirls, along the perimeter walls, and at the bottom. An EPA swirl regulator/concentrator is shown in Figure 4.10.

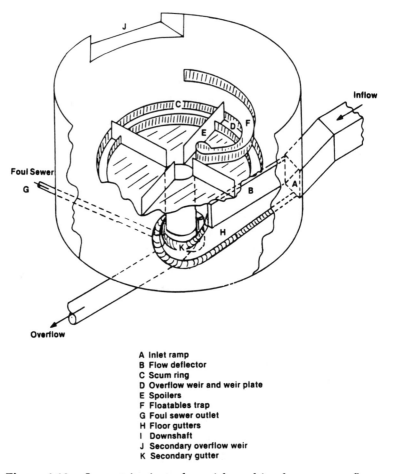

Figure 4.10. Isometric view of a swirl combined sewer overflow regulator/separator (U.S. EPA, 1982).

A Inlet ramp
B Flow deflector
C Scum ring
D Overflow weir and weir plate
E Spoilers
F Floatables trap
G Foul sewer outlet
H Floor gutters
I Downshaft
J Secondary overflow weir
K Secondary gutter

The swirl device can offer a compact unit that functions as both a regulator for flow control and a solids concentrator, and, when combined with treatment of the relatively heavy settleable solids, can provide an effective treatment system. There are a number of references (U.S. Environmental Protection Agency, 1973a, 1974b, 1977d, 1982, 1984b) that provide performance and design information for the EPA swirl regulator/concentrator. A degritter version of the EPA swirl has also been developed (U.S. Environmental Protection Agency, 1977d, 1981a) that has no underflow and removes only the grit portion.

STORAGE AND TREATMENT OPTIMIZATION

As stated previously, storage alone will offer only flow attenuation, and treatment alone will treat only a fraction of the stormwater flow or have such a large capacity to handle peak flows that the costs will be prohibitive. Therefore, combining the storage/treatment, finding the best balance, and, if possible, using existing facilities are likely to provide the most cost-effective solution for treatment of urban storm runoff.

No two situations are likely to be the same but a cost analysis to produce curves of storage alone, treatment alone, and the combined cost will produce an optimized cost curve, as shown in Figure 4.11. Factors such as the number of storms that are likely to exceed the capacity of the combined system need to be taken into account, but this approach will provide useful information on which to base a decision (U.S. Environmental Protection Agency, 1972b).

Due to the variable nature of storm events there will always be some storm events that generate runoff in excess of the storage/treatment capacity. The excess runoff will be discharged untreated to the receiving water. Use of the swirl regulator/concentrator described above can provide some treatment to the runoff that is either diverted to storage or the receiving water.

BENEFICIAL REUSE OF STORMWATER

The reuse of municipal wastewater for industry, nonpotable domestic usages, and groundwater recharge has been practiced for several years. In 1971 an EPA nationwide

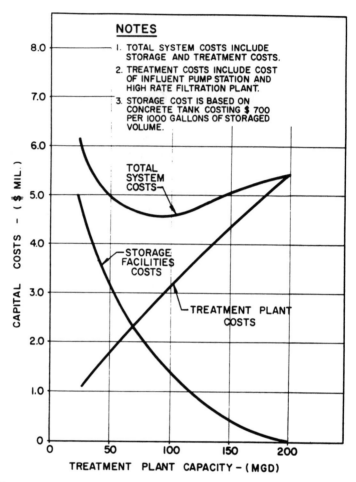

Figure 4.11. Estimated capital costs of storage and treatment for 200 MGD overflow (U.S. EPA, 1972b).

survey estimated that current reuse of treated municipal wastewater for industrial water supply, irrigation, and groundwater recharge was 53.5 billion gal/yr, 77 billion gal/yr, and 12 billion gal/yr (200 million m³/yr, 290 million m³/yr, and 45 million m³/yr), respectively (Environmental Protection Agency, 1975b). It is reasonable to expect that reuse of treated wastewater and/or stormwater for industrial cooling, nonpotable domestic water supply, and park and golf course irrigation will increase in the future.

Many of the treatments discussed above are likely to produce an effluent quality that is of a higher standard than that required to meet a stormwater permit. Where there are suitable circumstances, an opportunity exists to take advantage of this higher effluent quality for reuse of the storm runoff. The intended reuse will govern the level of treatment required, but careful selection, design, and use of pilot studies should result in the required combination of the above technologies to achieved required effluent quality.

The additional cost to provide treatment above that required to satisfy a discharge permit will need to be less than the cost of water from other sources, for economic viability. With increasing demands on potable water supplies, the concept of reuse, in particular where a nonpotable water quality standard is required, will make this an increasingly viable option. The chapter "Reclamation of Urban Stormwater" from *Integrated Stormwater Management* provides details and a hypothetical case study (Field et al., 1993).

REFERENCES

Bank, H. L., John, J., Schmehl, M. K., and Dratch, R. J., 1990. Bactericidal Effectiveness of Modulated UV Light. *Appl. Environ. Microbiol.* 56(12), 3888–3889.

Calderon, R. L., Mood, E. W., and Dufour, A. P., 1991. Health Effects of Swimmers and Non-point Sources of Contaminated Water. *Int. J. Environ. Health 1*, 21.

Camp Dresser and McKee, 1993. *State of California Storm*

Water Best Management Practice Handbooks. The California State Water Quality Control Board.

Colt, J., Tanji, K., Tchobanoglous, G., 1977. Impact of Dog, Cat, and Pigeon Wastes on the Nitrogen Budget of San Francisco Storm Runoff. Department of Water Science and Engineering University of California, Davis, No. 4015.

Field, R., 1990. "Combined Sewer Overflows: Control and Treatment." In: *Control and Treatment of Combined Sewer Overflows*, P. E. Moffa, ed., Van Nostrand Reinhold, New York, pp. 119–190.

Field, R., Forndran, A., and Dunkers, K., 1990. Demonstration of In-Receiving Water Storage of Combined Sewer Overflows: In a Marine/Estuarine Environment by the Flow Balance Method, Drainage Systems and Runoff Reduction. *Proceedings of the Fifth International Conference on Urban Storm Drainage*, Osaka, Japan, Ed. Y. Iwasa and T. Sueishi, Vol. 2, 759–764.

Field, R., O'Shea, M. L., and Chin, K. K., eds., 1993. *Integrated Stormwater Management.* Lewis Publishers, Boca Raton, FL.

Field, R., Pitt, R., Lalor, M., Brown, M. P., Vilkelis, W. V., 1994. Investigation of Dry-Weather Pollutant Entries into Storm Drainage Systems. *ASCE J. Env. Engg.,* October.

Field, R., and Struzeski, Jr., E. J., 1972. Management and Control of Combined Sewer Overflows. *J. Water Pollut. Control Fed.,* Vol. 44, No. 7, July.

Forndran, A., Field, R., Dunkers, K., and Moran, D., 1991. Balancing Flow for CSO Abatement. Water Environment Tech., Water Environment Fed. Vol. 3, No. 5, pp. 54–58.

Metcalf and Eddy, Inc., 1981. *Wastewater Engineering: Collection and Pumping of Wastewater.* McGraw-Hill Book Company, New York.

Metropolitan Washington Council of Governments, 1992a. *Analysis of Urban BMP Performance and Longevity.* Prince

George's County, Maryland, Department of Environmental Resources.

Metropolitan Washington Council of Governments, 1992b. *A Current Assessment of Urban Best Management Practices: Techniques for Reducing Non-Point Source Pollution in the Coastal Zone.* T. A. Schueler, P. A. Kumble, M. A. Heraty, MWCG, for U.S. EPA.

Metropolitan Washington Council of Governments, 1992c. *Design of Stormwater Wetland Systems: Guidelines for Creating Diverse and Effective Stormwater Wetlands in the Mid-Atlantic Region.* T. R. Schueler.

Metropolitan Washington Council of Governments, 1987. *Controlling Urban Runoff: A Practical Manual for Planning and Designing Urban BMPs.* T. R. Schueler, MWCG for Washington Metropolitan Water Resources Planning Board.

Minnesota Pollution Control Agency, 1989. *Protecting Water Quality in Urban Areas.*

New York State Department of Environmental Conservation, 1992. *Reducing the Impacts of Stormwater Runoff From New Development.*

O'Shea, M. L., and Field, R., 1992. "The Detection and Disinfection of Pathogens in Storm-Generated Flows." *Advances in Applied Microbiology,* Vol. 37, Academic Press. New York.

Pecher, R., 1969. The Runoff Coefficient and Its Dependence on Rain Duration. *Berichte aus dem Institut fur Wasserwirtschaft und Gesundheitsingenieurwesen,* No. 2, TU Munich (in German).

Pitt, R., 1989. *Source Loading and Management Model: An Urban Nonpoint Source Water Quality Model.* University of Alabama at Birmingham, AL.

Pitt, R., 1987. Small Storm Urban Flow and Particulate Washoff Contributions to Outfall Discharges. Ph.D. dissertation submitted to the Department of Civil and Environmental Engineering, University of Wisconsin, Madison.

Pitt, R., Ayyoubi, A., Field, R., and O'Shea, M. L., 1993. "The Treatability of Urban Stormwater Toxicants." In: *Integrated Stormwater Management*, R. Field, M. L. O'Shea, and K. K. Chin, eds., p. 121, Lewis Publishers, Boca Raton, FL.

Pitt, R., Barron, P., Ayyoubi, A., and Field, R., 1991. The Treatability of Urban Stormwater Toxicants. Proceedings 17th Annual RREL Hazardous Waste Research Symposium: Remedial Action, Treatment, and Disposal of Hazardous Waste, U.S. EPA, Risk Reduction Engineering Laboratory, Office of Research and Development, Cincinnati, OH, EPA/600/9-91/002 (NTIS PB 91 233 627).

Pitt, R., and Dunkers, K., October, 1993. "Lake Water Quality Improvements," from *Treatment of Stormwater Using the Flow Balancing Method*. Water Environment Federation, 66th Annual Conference & Exposition, Anaheim, CA, AC93-019-001.

Pitt, R., and Field, R., 1990. Hazardous and Toxic Wastes Associated with Urban Stormwater Runoff. Proceedings 16th Annual RREL Hazardous Waste Research Symposium: Remedial Action, Treatment, and Disposal of Hazardous Waste, U.S. EPA, Office of Research and Development, Cincinnati, OH, EPA/600/9-90-037 (NTIS PB 91 148 379).

Pitt, R. and McLean, J., 1986. *Toronto Area Watershed Management Strategy Study: Humber River Pilot Watershed Project*. Ontario Ministry of the Environment, Toronto, Ontario.

Roesner, L. A., Urbonas, B. R., and Sonnen, M. B., eds., 1989. Design of Urban Runoff Quality Controls. Proceedings of an Engineering Foundation Conference on Current Practice and Design Criteria for Urban Quality Control, American Society of Civil Engineers, New York.

Schmidt, S. D., and Spencer, D. R. 1986. The Magnitude of Improper Waste Discharges in an Urban Stormwater System. *J. Water Pollution Control Fed.*, Vol. 58, No. 7.

Soderlund, H., 1988. Recovery of the Lake Ronningesjon in Taby, Sweden; Results of Storm and Lake Water Treatment Over the Year 1981–1987. *Vatten*, Vol. 4, No. 44.

SCVWD, 1990. *Santa Clara Valley Nonpoint Source Study—Volume II: NPS Control Program.* Santa Clara Valley Water District, San Jose, CA.

Urbonas, B., and Stahre, P., 1993. *Stormwater: Best Management Practices and Detention.* Prentice-Hall, Englewood Cliffs, NJ.

U.S. Environmental Protection Agency, 1993a. *Handbook: Urban Runoff Pollution Prevention and Control Planning.* Metcalf and Eddy Inc., Office of Research and Development, Cincinnati, OH, EPA/625/R-93/004.

U.S. Environmental Protection Agency, 1993b. Urban Runoff Management Information/Education Products. Developed for U.S. EPA Region 5, Water Division, Wetlands and Watershed Section Watershed Management Unit and USEPA Office of Wastewater Enforcement and Compliance Permits Division, NPDES Program Branch, Stormwater Section, Version 1.

U.S. Environmental Protection Agency, 1993c. *Investigation of Inappropriate Pollutant Entries into Storm Drainage Systems: A User's Guide.* R. E. Pitt, M. Lalor, R. Field, D. D. Adrian, D. Barbé, EPA/600/R-92/238.

U.S. Environmental Protection Agency, 1992a. *Designing an Effective Communication Program: A Blueprint for Success.* R. M. Beech and A. F. Dake, University of Michigan, School of Natural Resources and Environment for USEPA Region 5.

U.S. Environmental Protection Agency, 1992b. *Storm Water Pollution Prevention for Construction Activities.* EPA-832-R-92-005.

U.S. Environmental Protection Agency, 1992c. *Effects and Treatment of Stormwater Toxicants.* Storm & Combined Sewer

Pollution Control Research Program, Edison, NJ, Cooperative Agreement CR819573.

U.S. Environmental Protection Agency, 1991a. Developing the Watershed Plan. In: *Nonpoint Source Watershed Workshop*, Cincinnati, OH, EPA/625/4-91/027 (NTIS PB92-137504).

U.S. Environmental Protection Agency, 1991b. Developing Goals for Nonpoint Source Water Quality Projects. In: *Nonpoint Source Watershed Workshop*, Cincinnati, OH, EPA/625/4-91/027 (NTIS PB92-137504).

U.S. Environmental Protection Agency, 1990. *Construction of a Flow Balancing Facility for Combined Sewer Overflow Retention, Project Enhancement and Floatables Removal.* EPA Region II, New York, Cooperative Agreement C361320.

U.S. Environmental Protection Agency, 1988. *Constructed Wetlands and Aquatic Plant Systems for Municipal Wastewater Treatment.* CERI, Cincinnati, OH, EPA/625/1-88/022.

U.S. Environmental Protection Agency, 1986. *Ambient Water Quality Criteria for Bacteria.* EPA-440/5-84-002.

U.S. Environmental Protection Agency, 1985. *Characterization, Sources, and Control of Urban Runoff by Street and Sewerage Cleaning.* R. E. Pitt, Consulting Engineer, Blue Mounds, WI, EPA-600/2-85/038 (NTIS PB 85-186500/Reb.).

U.S. Environmental Protection Agency, 1984a. *Health Effects Criteria for Fresh Recreational Waters.* A. P. Dufour, U.S. EPA, EPA-600/1-84-004.

U.S. Environmental Protection Agency, 1984b. *Swirl and Helical Bend Regulator/Concentrator for Storm and Combined Sewer Overflow Control.* W. C. Pisano et al., Environmental Design & Planning, Inc., Boston, EPA-600/2-8/116 (NTIS PB 85-102 523/Reb.).

U.S. Environmental Protection Agency, 1983a. *Evaluation of Catchbasin Performance for Urban Stormwater Pollution Control.*

G. L. Aronson et al., Environmental Design & Planning, Inc. Boston, EPA-600/2-83-043. (NTIS PB 83-217745).

U.S. Environmental Protection Agency, 1983b. *Health Effects Criteria for Marine Recreational Waters.* V. J. Cabelli, U.S. EPA, EPA-600/1-80-01.

U.S. Environmental Protection Agency, 1982. *Design Manual—Swirl and Helical Bend Pollution Control Devices.* R. H. Sullivan et al., American Public Works Association, Chicago, EPA-600/8-82/013 (NTIS PB 82-266 172).

U.S. Environmental Protection Agency, 1981a. *Field Evaluation of a Swirl Degritter at Tamworth, New South Wales, Australia.* G. J. Shelly et al., G. J. Shelly Consulting Engrs., Tamworth, NSW, Australia, EPA-600/2-81-063 (NTIS PB 81-219 602).

U.S. Environmental Protection Agency, 1981b. *Best Management Practices Implementation*, Rochester, New York. C. B. Murphy et al., O'Brien and Gere Engineers, Inc., Syracuse, NY, EPA-905/9-81-002 (NTIS PB 82 169 210).

U.S. Environmental Protection Agency, 1980. *Porous Pavement: Phase 1 Design and Operational Criteria.* E. V. Diniz, Espey, Hustn & Associates, Inc., Albuquerque, NM, EPA-600/2-80-135 (NTIS PB 81 104 138).

U.S. Environmental Protection Agency, 1979a. *Dual Process High-Rate Filtration of Raw Sanitary Sewage and Combined Sewer Overflows.* H. Innerfield et al., New York City Dept. of Water Resources, EPA-600/2-79-015 (NTIS PB 80-159 626/AS).

U.S. Environmental Protection Agency, 1979b. *Combined Sewer Overflow Abatement Program*, Rochester, NY–Vol. I, *Pilot Plant Evaluations.* F. J. Drehwing et al., O'Brien and Gere Engineers, Inc. Syracuse, NY, EPA-600/2-79-031b (NTIS PB 80-159 262).

U.S. Environmental Protection Agency, 1979c. *Screening/Flo-*

tation Treatment of Combined Sewer Overflows— Vol. II, *Full-Scale Operation*, Racine, WI. T. L. Meinholz, Envirex, Inc., Milwaukee, WI, EPA-600/2-79-106a (NTIS PB 80-130 693).

U.S. Environmental Protection Agency, 1979d. *Demonstration of Nonpoint Pollution Abatement Through Improved Street Cleaning Practices.* R. E. Pitt, Woodward–Clyde Consultants, San Francisco, EPA-600/2-79-161 (NTIS PB 80-108988).

U.S. Environmental Protection Agency, 1978. *Optimization and Testing of Highway Materials to Mitigate Ice Adhesion—Interim Report.* M. Kruker and J. C. Cook, Washington State University, Pullman, WA, EPA-600/2-78-56 (NTIS PB 280927/5).

U.S. Environmental Protection Agency, 1977a. *Urban Stormwater Management and Technology: Update and Users' Guide.* J. Lager et al., Metcalf and Eddy Inc., Palo Alto, CA, EPA-600/8-77-014 (NTIS PB 275 654).

U.S. Environmental Protection Agency, 1977b. *Catchbasin Technology Overview and Assessment.* J. Lager et al., Metcalf and Eddy, Inc. Palo Alto, CA, in association with Hydro-Research Science, Santa Clara, CA, EPA-600/2-77-051 (NTIS PB 270 092).

U.S. Environmental Protection Agency, 1977c. *Screening/Flotation Treatment of Combined Sewer Overflows—Vol. I, Bench-Scale and Pilot Plant Investigations.* M. K. Gupta et al., Envirex Environmental Science Div., Milwaukee, WI, EPA-600/2-77-69a (NTIS PB 272 834).

U.S. Environmental Protection Agency, 1977d. Swirl Device for Regulation and Treating Combined Sewer Overflows, EPA Technology Transfer Capsule Report. R. Field and H. Masters, U.S. EPA, Edison, NJ, EPA-625/2-77-012 (ERIC 2012, Cincinnati, OH).

U.S. Environmental Protection Agency, 1976a. *Development of a Hydrophobic Substance to Mitigate Pavement Ice Adhesion.*

B. H. Alborn and H. C. Poehlmann, Jr., Ball Brothers Research Corp., Boulder, CO, EPA-600/2-76-242 (NTIS PB 263 653).

U.S. Environmental Protection Agency, 1976b. *Proceedings of Workshop on Microorganisms in Urban Stormwater.* R. Field et al., U.S. EPA, Edison, NJ EPA-600/2-76-244 (NTIS PB 263 030).

U.S. Environmental Protection Agency, 1975a. *Bench-Scale High-Rate Disinfection of Combined Sewer Overflows with Chlorine and Chlorine Dioxide.* P. E. Moffa et al., O'Brien & Gere Engineers, Inc. Syracuse, NY, EPA-670/2-75-021 (NTIS PB 242 296).

U.S. Environmental Protection Agency, 1975b. *Current Municipal Wastewater Reuse Practices—Research Needs for the Potable Reuse of Municipal Wastewater.* C. J. Schmidt, EPA-600/9-75-007.

U.S. Environmental Protection Agency, 1974a. *Manual for Deicing Chemicals: Storage and Handling.* D. L. Richardson et al., Arthur D. Little, Inc., Cambridge, MA, EPA-670/2-74-033 (NTIS PB 236 152).

U.S. Environmental Protection Agency, 1974b. Relationship between Diameter and Height for Design of a Swirl Concentrator as a Combined Sewer Overflow Regulator. R. H. Sullivan et al., American Public Works Assoc., Chicago, EPA-670/2-74-039 (NTIS PB 234 646).

U.S. Environmental Protection Agency, 1974c. *Manual for Deicing Chemicals: Application Practices.* D. L. Richardson et al., Arthur D. Little, Inc., Cambridge, MA, EPA-670/2-74-045 (NTIS PB 239 694).

U.S. Environmental Protection Agency, 1974d. *Combined Sewer Overflow Treatment by the Rotating Biological Contactor Process.* F. L. Welsh and D. J. Stucky, Autotrol Corp., Milwaukee, WI, EPA-670/2-74-050 (NTIS PB 231 892).

U.S. Environmental Protection Agency, 1973a. *The Dual-*

Function Swirl Combined Sewer Overflow Regulator/Concentrator. R. Field, U.S. EPA, Edison, NJ, EPA-670/2-73-059 (NTIS PB 227 182/3).

U.S. Environmental Protection Agency, 1973b. *Combined Sewer Overflow Seminar Papers.* U.S. EPA Storm and Combined Sewer Research Program and NYS-DEC, EPA-670/2-3-077 (NTIS PB 235 771).

U.S. Environmental Protection Agency, 1973c. *Physical–Chemical Treatment of Combined and Municipal Sewage.* A. J. Shuckrow et al., Pacific NW Lab, Battelle Memorial Institute, Richland, WA, EPA-R2-73-149 (NTIS PB 219 668).

U.S. Environmental Protection Agency, 1972a. A Search: New Technology for Pavement Snow and Ice Control. D. M. Murray and M. R. Eigermann, ABT Associates, Inc., Cambridge, MA, EPA-R2-72-125 (NTIS PB 221 250).

U.S. Environmental Protection Agency, 1972b. *High-Rate Filtration of Combined Sewer Overflows* (Cleveland). R. Nebolsine et al., Hydrotechnic Corp., New York, EPA-110EY104/72 (NTIS PB 211 144).

U.S. Environmental Protection Agency, 1971. *Environmental Impact of Highway Deicing.* Edison Water Quality Research Laboratory, Edison, NJ, EPA11040GKK06/71 (NTIS PB 203 493).

U.S. Environmental Protection Agency, 1970a. *Combined Sewer Regulator Overflow Facilities.* American Public Works Association, Chicago, 11022DMU07/70 (NTIS PB 215 902).

U.S. Environmental Protection Agency, 1970b. *Combined Sewer Regulation and Management—A Manual of Practice.* American Public Works Association, Chicago, 11022DMU08/70 (NTIS PB 195 676).

Viessman JR., W., Knapp, J., and Lewis, G. L., 1977. *Introduction to Hydrology.* 2nd edition, p. 69. Harper and Row, New York.

Walesh, S. G., 1989. *Urban Surface Water Management*. Wiley, New York.

Wanielista, M. P., and Yousef, Y. A., 1992. *Stormwater Management*. Wiley, New York.

Water Environment Federation, 1992. *Design of Municipal Wastewater Treatment Plants. Volumes I and II, WEF Manual of Practice*, No. 8, Alexandria, VA. ASCE *Manual and Report on Engineering Practice* No. 76, 2nd ed., ASCE, NY.

Water Environment Federation, 1990. *Natural Systems for Wastewater Treatment: Manual of Practice*, FD-16.

Yu, S. L., Kasnick, M. A., and Byrne, M. R. 1993. "A Level Spreader/Vegetative Buffer Strip System for Urban Stormwater Management." In: *Integrated Stormwater Management*. R. Field, M. L. O'Shea, and K. K. Chin, eds., p. 93. Lewis Publishers, Boca Raton, FL. Camp Dresser & McKee, 1993.

5

Case Histories

INTRODUCTION

This chapter presents six case histories:

1. a stormwater management utility, in the far west, that represents more than 20 years of operating experience

2. a typical Industrial Stormwater Pollution Prevention Plan

3. an innovative stormwater recycling program to eliminate typical end-of-pipe stormwater discharges

4. a comparison of different cost-allocation methodologies and an example of a wet-weather fee structure that relates more closely to actual urban runoff impacts and associated abatement costs

5.   an industrial stormwater case that involved field calculations, system and receiving-water modeling, and design and construction of an industrial stormwater abatement facility.

6.   a case study comparing annual pollutant loadings from loadings from urban stormwater to CSO discharges in a lake watershed in the northeast.

1. Bellevue, Washington's Storm and Surface Water Utility

Case History by Wendy Skony, City of Bellevue, Washington

Bellevue, Washington (see Figure 5.1), the fourth-largest city in the state, is a city known for its water. The rainy climate, two adjacent relatively large freshwater lakes (Lake Sammamish and Lake Washington), 11 major drainage basins, 740 acres of wetlands, 50 miles of open streams, three small lakes, and numerous ponds make water management a necessity for the city of Bellevue.

Rapid growth and development began to take a toll on the environment by the late 1960s. Increased impervious area accelerated surface water runoff, thus creating flood problems, water pollution, erosion, and degraded salmon runs. At that time, stormwater management was a responsibility of the Public Works Department, as the city had no comprehensive plan for controlling runoff. Stormwater was regarded as a temporary problem that did not warrant enough funds for a capital improvement program. The city council and community leaders began to recognize the need for a comprehensive approach to stormwater management when local citizens voiced their concern about the degradation of the city's streams, wetlands, and open spaces.

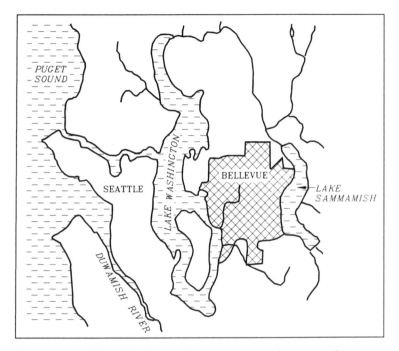

Figure 5.1. City of Bellevue, Washington, and surroundings.

In 1965, the Revised Code of Washington was changed to allow for the establishment of utilities for stormwater control. During the next five years, a series of events led to the formation of Bellevue's Storm and Surface Water Utility. The city council authorized a study of Bellevue's streams and appointed a citizen's committee to study and recommend standards and procedures for preserving these streams, as described in Figure 5.2. A clearing and grading ordinance, along with a resolution establishing a stream-preservation policy, was passed. In addition, a bond issue that included funds for a new storm drainage study was passed.

On March 7, 1974, following a public hearing, the city council adopted an ordinance allowing the formation of the region's first Storm and Surface Water Utility. Revenues grew slowly from a one-time contribution from the

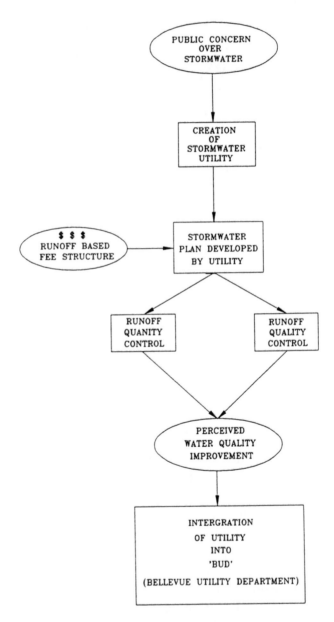

Figure 5.2.   Evolution of the Bellevue Stormwater Utility and BUD, the Bellevue Utility Department.

City's general fund, until rates were raised to fund the Capital Improvement Program (CIP). Three major rate increases occurred in 1980, 1982, and 1986, which were approximately 70%, 90%, and 35%, respectively. Since then, rate increases have been in the single digits, to cover inflation. Any funds not used at year-end are carried forward to the next year.

Rates are based on the estimated amount of runoff that individual properties contribute to the total drainage system. This method of billing avoided the problem of "bottom of the hill" versus "top of the hill." Although those properties in the upland areas perceived no direct benefit, the owners were still required to pay the fee, as runoff from the "top" contributed to the problem at the "bottom of the hill." Likewise those property owners at the bottom of the hill received a direct benefit from reduced property damage, etc., but they did not have to pay for all the improvements. Each property is classified by a runoff coefficient, based on its percentage of developed property. The classification, combined with total property area, determines the service charge that is billed every two months. Billing-system data are continually updated as property owners change and undeveloped land becomes divided and developed.

The basic concept underlying all the policies concerning stormwater management in Bellevue is the use of a natural surface-water drainage system to provide conveyance and disposal of runoff without degrading the system. This system uses regional, in-stream flood control facilities to attenuate peak flows for older developments. The drainage-utility system consists of an integrated network of pipes and stream channels as the conveyance system, and lakes, wetlands, ponds, and detention basins for flow equalization and water-quality control. On-site flood controls are required of new industry and commerce, thereby placing the burden of mitigating environmental impacts on the developer rather than the general public. The open-stream concept was less costly than traditional storm sewers and has proven to be more environmentally sensitive to hydrologic ecosystems.

Utility programs include operations and maintenance, administration, water quality, Capital Improvement Program, and public education. Flood control is one of the most successful utility programs. Water levels, stream flow, and rainfall are all monitored via computer at Bellevue's eight major flood control facilities. Raging flood waters are held back by remote control and released at a low rate. On a weekend or after hours, the computer calls employees at home if stormwater levels become a concern. With successful flood control systems in place, the utilities focus has shifted to water quality.

Although the Storm and Surface Water Utility was a separate department for 10 years, in 1993 it joined the city's other utilities to become one comprehensive Bellevue Utilities Department (BUD). The Storm and Surface Water Utility, the Water and Sewer Utilities, and the Solid Waste Program fall under the utility umbrella, affording a greater environmental focus for all. Many educational and maintenance programs are now accomplished with greater efficiency, as redundancy has been reduced, thus providing enhanced customer service at a lower cost.

Bellevue's Storm and Surface Water Utility, currently BUD, marked its twentieth anniversary in March, 1994. By creating a system for surface water management and making the preservation of the environment a priority, Bellevue is still known as a well-planned, nature-oriented city. The Utility, through which the funding and planning have taken place over the past 20 years, may serve well as a model for communities with similar stormwater management problems.

2. ACME PLATING STORMWATER POLLUTION PREVENTION PLAN

Case History by HRP Associates, Inc.

Introduction

"Acme Plating" (Connecticut) is a plater of precious metals for electronics and other related industries. (The name "Acme Plating" is used here to protect the confidential interest of the actual company.) The site is approximately 2.0 acres in size and lies approximately 1000 feet east of the Acme River. The on-site facilities include the main 36,500-square-foot production and office building, a 600-square-foot metal hydroxide filter cake storage shed, and a small, 480-square-foot, chemical storage shed. The majority of the site drains in a sheet-flow manner to the on-site catch basins.

Acme Plating is covered by the Connecticut Department of Environmental Protection's (DEP) General Permit for the Discharge of Stormwater Associated with Industrial Activity, by virtue of the company's placement in Standard Industrial Classification (SIC) Code Number 3471 (plating and polishing), and due to storm water discharge from the site via a point source to the Acme River. A Stormwater Pollution Prevention Plan (SPPP) has been developed and includes the following conditions:

implementation of SPPP recommendations

sampling and analysis

employee training

periodic site inspections

plan update.

An overview of the primary characteristics of the SPPP is provided below.

Pollution Prevention Team

The purpose of the Pollution Prevention Team is to establish which specific individuals will be responsible for implementation, maintenance, and revision of the SPPP. Each member's responsibilities are summarized in Figure 5.3. The Team Coordinator is the site contact for any

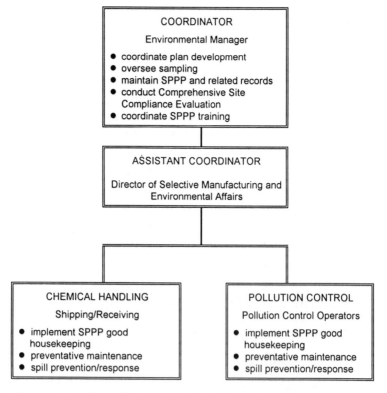

Figure 5.3.   Acme Plating Pollution Prevention Team

inquiries regarding implementation of the facility's SPPP and is responsible for keeping the Plan current.

Inventory of Exposed Material

An inventory of materials at the Acme Plating facility that are exposed to stormwater and the location of each exposed material have been quantified. Also identified are materials management practices currently employed by Acme Plating to minimize contact of materials with stormwater.

The primary outdoor structural measure that prevents or reduces pollutants in the stormwater running off the site is the maintenance of materials under cover. An additional structural measure is the presence of sumped catch basins in the storm drain system that allow for some limited soilds settling. The primary nonstructural measure is the regular inspection of all outdoor areas, including roof areas near exhaust vents. These areas are inspected weekly, except for the roof, which is inspected every four weeks.

Inside the facility there are a number of structural measures to prevent potential pollutants from entering stormwater. The chemical drums are stored inside concrete bermed areas. There is a trench system around the production areas to divert any spills to the wastewater treatment system. There is also a 60-gallon sump in the chemical storage area to collect any spills.

Some of the waste materials in the inventory are contained in portable roll-offs and dumpsters. These container locations may vary from time to time to accommodate site operations and activities. Short-term or temporary variations are consistent with the SPPP as long as the controls and measures described herein are maintained.

No spills and/or leaks of five gallons or more of toxic or hazardous substances that have been exposed to stormwater runoff have occurred since October 1, 1989.

Summary of Potential Pollutant Sources

The following narrative describes potential pollutant sources at specific facility areas. It also identifies specific pollutants that could potentially impact stormwater runoff quality.

Loading Docks
The facility has one loading dock. There is potential for oil and metal pollutants to enter stormwater at the dock, in the form of oils, which coat some incoming parts, and heavy metals, which could leach from parts if they experience long-term exposure to acidic rainfall. Potential chemical pollutants at the dock include all chemical compounds used at the facility. These include acids, alkalis, and solvents. An accidental uncontrolled spill during delivery is the major potential pollution pathway.

Roof Areas
There are numerous vents on the Acme Plating roof that have the potential to emit chemicals that may result in a residual deposit on the surrounding roof area and subsequently lead to runoff. These potential sources are located over the production area of the main building. Following is a listing of the primary chemicals of concern and their major sources.

Chemical	*Source*
Lead	Plating
Copper cyanide	Plating
Potassium cyanide	Plating

Sodium	Plating
Hydrochloric acid	Plating and stripping
Nitric acid	Plating

Dust-Generating Processes

The only significant potential source of dust generation is the metal particles from the buffing operations. Escape of particles with stormwater could result in increased metals levels in the runoff. This area drains directly to the Acme River via the permitted stormwater outfall.

Outdoor Storage of Raw Materials

Raw materials stored outdoors at Acme Plating include both gaseous and liquid-phase chemicals. Gaseous chemicals are not a potential pollutant source of the type under consideration because any accidental escape would be to the atmosphere and not to the ground surface. The liquid chemicals, acids, caustics, and solvents, are potential sources of decreased or increased pH pollution or of halogenated organics, should a spill or leak occur.

Outdoor Storage of Waste Materials

Waste materials stored out of doors include paper, cardboard, and other MSW (municipal solid waste), and dewatered sludges. The MSW is not a significant source of chemical pollutants. The sludges are potential sources of metals and cyanide.

Bulk Delivery Activities

Deliveries of sodium hypochlorite are a source of increased pH, should a release occur.

Soil Erosion

Significant soil erosion results in increased suspended solids in stormwater runoff. There are no significant on-site sources of potential soil erosion.

Measures and Controls

The SPPP discusses measures and controls that will minimize pollutant generation and transport at each potential pollutant source identified previously. Many measures and controls described below are already being conducted by Acme Plating staff.

Physical Measures and Controls
   Stormwater management controls appropriate for this facility include the following physical elements.

Good Housekeeping   Good housekeeping consists of maintaining a clean, orderly facility to prevent unnecessary rain water contact with materials that may result in stormwater contamination. No washing of vehicles or equipment outdoors that would allow wash waters to enter the storm drainage system is allowable by the General Permit.

Preventive Maintenance   Preventive maintenance at this site will include maintenance of systems that include piping and valves or container storage. It will also consist of periodic site inspections to identify potential pollutant sources as well as inspections of the stormwater drainage system to ensure its proper operation.

Spill Prevention and Response Procedures   Spill prevention recomendations for each of the potential pollutant sources have been developed. Spill response procedures enacted under this SPPP should be conducted in a manner consistent with the Emergency Response Plan prepared in accordance with OSHA Regulation 29 CFR 1910.120 under separate cover.

Chemical Storage Secondary Containment   The general permit requires that all chemicals be stored in areas provided with impermeable containment that will hold at least the volume of the largest chemical container, or 10% of the total volume of all containers in the area, whichever is larger, without overflow from the containment area.

Runoff Management

For purposes of this Plan, *runoff management* is defined as those practices that divert, infiltrate, reuse, or treat stormwater runoff, and is distinct from those practices that limit exposure of potential pollutants to direct rainfall or runoff.

Runoff management at the Acme Plating facility consists of the use of sumped catch basins that allow for solids settling if maintained clear of debris. Should annual stormwater sampling and analysis indicate a long-term problem of stormwater pollution, it may be appropriate to consider additional management practices.

Visual Inspections

Currently, the Acme Plating staff conducts regular inspections of the entire site, including buildings, roof areas, storage sheds, and outdoor areas. The site is inspected weekly except for the roof, which is inspected monthly. Written inspection reports and written reports of follow-up actions should be maintained at the facility.

Comprehensive Site Compliance Evaluation

A Comprehensive Site Compliance Evaluation (CSCE) must be conducted by qualified personnel at least twice per year to evaluate the effectiveness of the SPPP and to identify any needed Plan revisions. The CSCE will be most effective if performed in the spring and the fall, with the actual inspection conducted during a rainfall event.

Two elements comprise the CSCE:

• a visual inspection of all site features as well as any equipment necessary for proper maintenance,

repair, and execution of the measures and controls identified in the SPPP
- a written report summarizing the evaluation, and including:
  - names of personnel conducting inspection
  - date of inspection
  - observations of actions needed
  - resolution of needed actions
  - signature of Acme Plating owner.

The CSCE reports must be maintained on-site for at least five years.

It is a requirement of the General Permit that the SPPP be kept current by means of any necessary revisions that are identified through the semiannual CSCE process. In addition, any modification of measures and controls necessitated by these revisions must be implemented within 60 days of CSCE completion. The SPPP should be amended by Acme Plating whenever:

- there is a change at the site that has an effect on the potential to cause pollution of state waters; or
- the actions required by the SPPP fail to ensure or adequately protect against pollution of state waters.

Employee Training

The employee training plan for Acme Plating will consist of two basic programs, which will include all employees, as listed below. The primary purpose of the training is to make all employees aware of the need to prevent pollutants from entering stormwater runoff. Most employees' responsibilities, which should be communicated to them through the training, are limited to observing and reporting to their direct supervisors conditions that may lead to exposure of pollutant sources to direct rainfall or rainfall runoff.

Comprehensive Training Plan

Comprehensive Training is designed for those employees whose regular work duties are directly or indirectly involved with industrial activities exposed to stormwater. This training should consist of a 30–60-minute session, at which the following items are discussed:

A. an overview of the General Permit, including the Pollution Prevention Team organization as well as potential fines for non-compliance

B. details of the Stormwater Pollution Prevention Plan, including potential pollutant sources, measures and controls, and spill prevention and response

C. Comprehensive Site Compliance Evaluation.

Limited Training Plan

Limited Training is designed for production and administrative employees. This training should consist of a 15–30-minute review of the General Permit and its basic requirements. Section A., General Permit Overview, of the Comprehensive Training Plan, can be used as the basic outline for the Limited Training Plan. Further, a brief explanation of the General Permit and the importance of all employees' awareness and reporting responsibilities should be included as part of new employees' orientation.

Training Evaluation

A written evaluation of performance will be done by the Pollution Prevention Team Leader immediately after a drill or actual spill clean-up. In the event that exercises show that employee performance must be updated, additional training will be provided. The stormwater pollution prevention training will be updated when new hazards are introduced or when procedures have been updated or revised.

Monitoring and Reporting

Parameters to Be Monitored

Annual stormwater monitoring is required by the General Permit to be conducted annually beginning in 1993. Parameters for the analyses that are required are as follows.

A.   Standard stormwater parameters

Total oil and grease (mg/l)
pH
Chemical oxygen demand (mg/l)
Total suspended solids (mg/l)
Total phosphorus (mg/l)
Total Kjeldahl nitrogen (mg/l)
Nitrate as nitrogen (mg/l)
Fecal coliform (#/100 ml)
Total copper (mg/l)
Total zinc (mg/l)
Total lead (mg/l)

B.   Parameters required to be monitored in Acme Plating SPDES Permit (also required for stormwater analysis)

Cadmium	Total residual chlorine
Chromium, hex	Phenols
Chromium, total	Cyanide, amenable
Iron	Cyanide, total
Nickel	Fluoride
Tin	Method 601 (halogenated organic compounds)
Silver	Method 602 (aromatic organic compounds)
Gold	
Aluminum	

C.   Parameters limited in the EPA Effluent Guideline contained in 40 CFR Part 400–499

All parameters are included in Sections A and B above.

Stormwater Monitoring Procedures

Stormwater Sampling Instructions include and relate to the following items.

- event selection
- general procedures
- sample and data collection procedure
- sample containers
- sample submission.

The procedures should be reviewed each year before sampling and analyses are completed. The sampling location at the Acme Plating facility is the last catch basin in the storm drainage system before runoff leaves the site.

Retention of Records

For each measurement or sample taken, Acme Plating must record the following information:

- the place, date, and time of sampling
- the person(s) collecting samples
- the dates and times the analyses were initiated
- the person(s) or laboratory that performed the analyses
- the analytical techniques or methods used
- the results of all required analyses.

All records and information resulting from the monitoring activities required by the General Permit, including all records of analyses performed and any applicable calibration and maintenance of instrumentation, must be retained for a minimum of five years following the expiration of the Permit, or longer if required by the DEP. At this time the General Permit expiration date is October 1, 1997.

Table 5.1. Freshwater Acute Toxicity Numerical Water Quality Criteria.

Substance	Criterion
Copper	0.0143 mg/l
Lead	0.034 mg/l
Zinc	0.0353 mg/l
Cadmium	0.0018 mg/l
Chromium, hex	0.016 mg/l
Chromium, total	0.980 mg/l
Nickel	0.790 mg/l
Silver	0.0012 mg/l
Cyanide	0.022 mg/l
Chlorine	0.019 mg/l

Reporting Requirements

1. All storm water monitoring results must be retained on-site and made available to DEP upon request.

2. If any stormwater sample results exceed the CT water quality standards listed below, Acme Plating is required to perform 24-hour acute toxicity biomonitoring testing on the next stormwater sample submitted for analysis. If additional parameters are required for stormwater analysis at this facility in the future, their associated acute toxicity criteria should be listed in Table 5.1.

3. If any 24-hour acute toxicity biomonitoring test results in greater than 50% mortality to either test species during a valid test, Acme Plating must submit a report to the DEP including the following information:

a. the complete results of the acute toxicity biomonitoring test, including the percent survival in each replicate test chamber and any physical/chemical monitoring of test solutions conducted prior to, during, or upon completion of the test

b. results of any chemical analyses conducted on samples of stormwater believed to be representative of the samples used in the acute toxicity biomonitoring test

c. a copy of the discharge registration form for the discharge.

4. If Acme Plating monitors any pollutant at the designated sampling location more frequently than required by the permit, using approved analytical methods, the results of such monitoring must be retained and reported as outlined above.

Other Discharge Requirements
The General Permit contains the following additional discharge requirements.

1. There should be no distinctly visible floating scum, oil, or other matter contained in the stormwater discharge. Excluded from this are naturally occurring substances such as leaves and twigs, provided no person has placed such substances in or near the discharge.

2. The stormwater discharge should not result in pollution due to acute or chronic toxicity to aquatic and marine life, impair the biological integrity of aquatic or marine ecosystems, or result in an unacceptable risk to human health.

3. WYMAN-GORDON STORMWATER RECYCLING PROGRAM

Case History by Woodard & Curran, Inc. (Source: *Industrial Wastewater*, Vol. 2, No. 5 © 1994 Waste Environment Federation; reprinted with permission.)

The Wyman–Gordon Company is a metals forging facility located on a 189-acre manufacturing complex in North Grafton, Massachusetts. This facility houses several metalworking operations, including forging, heat treating, chemical etching, and grinding. Wyman–Gordon uses lubricating and cutting oils, greases, coolants, and acids in its forging and related processes. As a result of EPA stormwater regulations, Wyman-Gordon opted to recycle their stormwater and pretreated process wastewaters in a Runoff Management Facility (RMF) rather than provide conventional stormwater treatment.

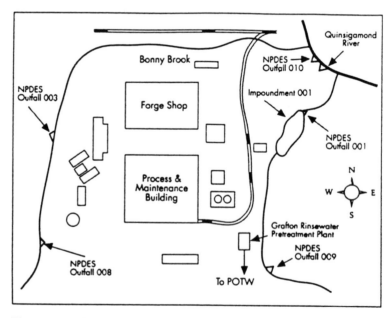

Figure 5.4.   Wyman–Gordon Co. Facility Plan

Previously, stormwater runoff from the building roof and yard areas in the northern, eastern, and northwestern portions of the site flowed by gravity to an existing impoundment (001) and was discharged at an NPDES-permitted outfall (001). Impoundment 001 also received some pretreated process wastewater, fire-protection water and noncontact cooling water. Runoff from other areas was collected and discharged at four other on-site NPDES-permitted outfalls, as shown in Figure 5.4.

Instead of incorporating expensive advanced treatment processes into the treatment system, Wyman-Gordon decided to institute a stormwater recycle program to eliminate the regular discharge at Outfall 001 and reduce the amount of water withdrawn from process water supplies. This decision was based on years of evaluation of site-specific stormwater concerns and EPA regulations.

Designing the RMF

The final design began with the development and documentation of design criteria. A design basis memorandum (DBM) provided a mechanism for obtaining Wyman-Gordon's approval and was used to initiate early discussions with the regulatory agencies that would permit the project. The DBM represented approximately the 30% design level and included design flows and loads; a preliminary process flow diagram and hydraulic profile; a site plan; criteria for the process, structural and civil designs; a cost opinion; and a final design schedule.

The 25-year, 24-hour design storm was selected to design the stormwater treatment facilities. The RMF effluent is intended to reduce the use of river water as a process source. With TSS, oil and grease, and pH identified as the parameters of concern, the expected recycle water quality, as found in the river water, were as follows:

TSS	<1–10 mg/l
Oil and grease	<1 mg/l
pH	5.0–8.0

Design flows and pollutant loadings were established to size the necessary process equipment.

Because the RMF would receive both stormwater runoff and some pretreated process wastewaters, dry- and wet-weather flows had to be considered. A water balance was performed to represent the flow of water before and after the recycle system began operating.

The water balance representing the "before-recycle" scenario is presented in Figure 5.5. Process water sources include on-site wells, the city water supply, and river water. Discharges from the site include NPDES outfalls, rinse water pretreatment plant effluent, sanitary wastewater, and other losses.

The "after-recycle" water balance shows that all stormwater runoff can be used as process water makeup (Figure 5.6), and that the system will reduce both the rate at which water is withdrawn from the river and the

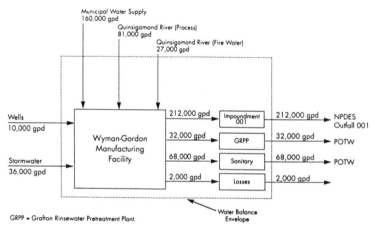

Figure 5.5. Before-recycle water balance

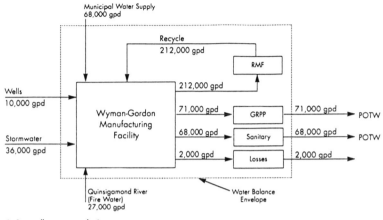

Figure 5.6. After-recycle water balance

amount of city water needed for process makeup (68,000 vs. 160,000 gallons per day).

The RMF was designed to serve three storm drain lines and effluent from a new oily-wastewater pretreatment system that will remove free and emulsified oils from press waters. A junction chamber was designed to intercept and reroute flows without throttling the storm drainage system or allowing solids to settle out. The sedimentation basin was sized to retain the design storm volume and process water flows, provide adequate surface area and retention time for solids settling, and serve as an equalization tank.

The RMF unit processes are shown in Figure 5.7 and described below. A layout of the RMF site is shown in Figure 5.8.

Grit Chamber

The grit chamber was designed to lower the influent flow velocity, thereby allowing for the removal of particles with a specific gravity greater than 2.65 and of a size that would be retained on 100-mesh screen. A slotted pipe

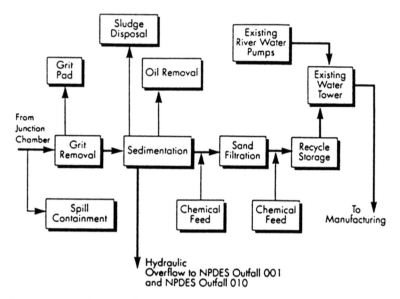

Figure 5.7.   Process flow diagram

skimmer in the grit chamber will drain the skimmed oil and water mixture to a sump in the RMF building, where the oil will be removed. The grit chamber will also provide added storage capacity (188,000 gallons).

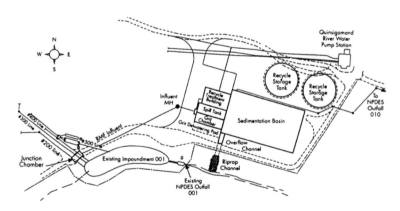

Figure 5.8.   Runoff management facility site layout

Sedimentation Basin

The sedimentation tank is an inground concrete tank open to the air and divided by a center wall. Sized to contain the design storm runoff volume of 2.06 million gallons and allow for 2 feet of freeboard, the tank has a total capacity of 2.34 million gallons. Its size (180 feet long, 100 feet wide, and 17 feet deep) and configuration provide enough surface area and retention time to allow additional settling of solids.

Oil Skimming

Oil skimming will be performed in two locations: the grit chamber and a baffled sump inside the RMF building. A floating-tube-type oil skimmer will remove floating oils, which will be stored temporarily for use as fuel in a specially permitted waste-oil burner in the RMF building. In Massachusetts, waste oil is considered a hazardous waste, so secondary containment was required for 100% of the storage tank volume.

pH Adjustment

Because of the acidic nature of rainfall in the Northeast (pH levels as low as 5.8 have been measured in the existing Impoundment 001), provision for pH adjustment were included in the RMF design. Sodium carbonate may be fed into the line as wastewater is pumped from the sedimentation tank to the sand filters. Besides meeting process water needs, a more neutral pH will meet the NPDES permit limit range of 6.5 to 8.0 through Outfall 010 and 001.

Sand Filtration

Filtration will be achieved by two Parkson upflow sand filters. These continuous-backwash systems eliminate the need for large volumes of backwash water (backwash flow is approximately 5 to 7% of the forward flow).

Recycle Water Disinfection

Recycling treated runoff could result in increased BOD concentrations in the recycle water and create a biofouling problem in the process water distribution system. To prevent this from becoming a problem, a sodium hypochlorite feed system was included in the design.

Storage and Distribution

Recycled water will be stored in two glass-fused-to-steel above-ground storage tank (ASTs) with aluminum domes, each with a capacity of 1.5 million gallons. Recycle water will be pumped, as needed, from the ASTs to the process water distribution system. If there is insufficient recycle water in storage, existing pumps will receive a signal to provide supplemental water from the Quinsigamond River.

Wyman-Gordon's innovative management approach to its stormwater runoff is expected to assure consistent compliance with EPA's stormwater regulations while reducing its dependence on outside water sources.

4. MANCHESTER, NEW HAMPSHIRE, WET-WEATHER FEE STUDY

Case History by Camp Dresser & McKee, Inc.

Manchester, New Hampshire, is a city of approximately 95,000 located in southern New Hampshire. Since 1992, the city has been developing a CSO facilities plan intended to reduce the nearly 50 combined sewer overflows that occur annually in the city. As the city has evaluated appropriate technical solutions, it has also been concerned with developing the most equitable means of allocating the costs of the control program among its customers. The city's goal is to bill customers in proportion to the burden they impose on the system.

Communities with wet-weather control programs, either for stormwater or combined flows, have traditionally allocated costs to customers by using two primary rate methods: general tax revenues and sewer use fees. With general tax systems, the amount a customer is billed varies with the value of the property owned by the customer. A customer with a very valuable parcel will pay more than someone with a less valuable parcel. Alternatively, sewer user fee rate systems allocate costs to customers based on relative wastewater flows, typically estimated through metered water consumption. A customer using 200,000 gallons of water per year will normally pay twice as much as someone consuming 100,000 gallons of water.

Manchester is evaluating whether either of these rate systems equitably allocates the costs of the CSO control program and whether a more equitable alternative exists. One option being evaluated is a user fee based on the amount of impervious area on a customer's property, similar to that used by the Bellevue Utilities Department and nearly 65 other communities that have established stormwater utilities. Under this type of system, "the more you pave, the more you pay." This type of allocation method assumes that the amount of use a customer gets from the wet-weather control system is directly related to the amount of impervious area on the customer's land.

Figure 5.9 illustrates the relationship between the amount of runoff generated from a parcel and the amount of impervious cover on the parcel, using empirical data from the National Urban Runoff Program. These data support the notion that impervious area is a good indicator of runoff and, thus, of demands on the control system. (Although this figure plots water quantity, a similar relationship exists for water quality.) The amount of runoff generated by a single parcel increases as the amount of

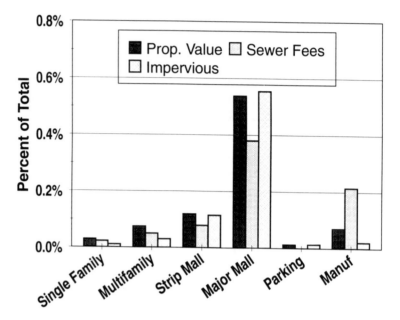

Figure 5.9.   The relationship between runoff and impervious cover.

impervious cover on that parcel increases. The amount of runoff generated from a single-family residential parcel is nearly double the amount generated by the same-sized parcel in its natural state. A multi-family residential parcel or office park generates over three times the runoff, and a regional shopping mall seven times the amount of runoff.

Given the relationship between runoff and impervious area, the next question is: Which rate allocation method best matches customer demand? Based on data developed for Manchester, New Hampshire, Figure 5.10 compares the impact of alternative cost allocation methodologies on several customer types. This figure compares the relative bills for various customer types, based on the three billing methodologies mentioned herein. As an example, 100 single-family residential parcels, on average, represent less than 4/100 of a percent of the total assessed value in the

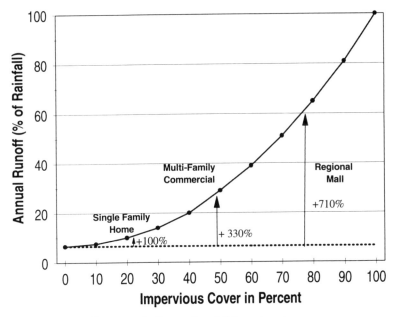

Figure 5.10. Impact of alternative billing structures on customer types.

city and consume approximately 3/100 of a percent of annual water demand, but contribute less than 1/100 of a percent of total wet-weather flow. This clearly suggests that either property tax or sewer use systems shift more costs to single-family residential properties than their collective share of wet-weather flows. Several analogous points are listed below.

- Major regional malls and strip malls would pay significantly less under a water consumption system than they would using either property values or impervious area. However, if the billing base is property values, these customers will still pay less than their contribution to the problem. (This is typical of the pattern for a wide range of commercial properties, including office buildings and some office parks.)

- Manufacturing facilities, at least in this particular data set, would pay significantly more than their contribution to the problem, if either property value or water consumption is used. (Frequently, manufacturing firms are overassessed based on property value, but underassessed based on water consumption unless they use a wet manufacturing process.)
- Facilities, like parking lots and warehouses, demonstrate most dramatically the inequity of the property value and water consumption methodologies. Both have a relatively low property value, and negligible water use, but contain large impervious areas.

Figure 5.10 demonstrates that Manchester's general tax revenue and sewer use fee systems do not allocate costs to customers proportionate to customers' demands on the wet-weather system. This is typical of most communities and explains why a large number of communities have adopted stormwater utility user charges based on impervious area.

5. Industrial Stormwater Case Study

Case History by Moffa & Associates

Overview

The following case history represents an actual project involving stormwater and receiving-water characterization, establishment of wet-weather standards, and design and construction of an industrial stormwater treatment facility. Due to the confidential nature of the site, any data representative of the actual site have been modified for illustrative purposes. The approach illustrated below has been accepted by the state regulatory agency and endorsed as an example methodology.

Background

The stormwater from a large industrial complex discharges to a small stream. The dry-weather flow from the facility consists largely of ground-water infiltration. The wet-weather flow is collected from an extensive network of roof drains and catch basins throughout the facility. Upon renewal of an existing State Pollutant Discharge Elimination System (SPDES) permit, the facility was subject to effluent discharge limitations for a pollutant XYZ. During the first several months of compliance sampling, the industry's discharge permit was violated on several occasions.

A methodology was developed to determine the treatment-system capacity required to meet water quality standards in the receiving stream over a range of rainfall conditions. Monitoring data, combined with mathematical modeling, served to produce a basis of design for treatment. Additionally, continuous simulation was used to identify the annual reduction in pollutant XYZ resulting from a chosen treatment alternative. A flowchart of this process is provided in Figure 5.11.

Data Collection

The monitoring and sampling program consisted of continuous rain gaging, flow monitoring, and XYZ sampling at the discharge point and in the receiving stream.

The rainfall and flow monitoring programs were con-

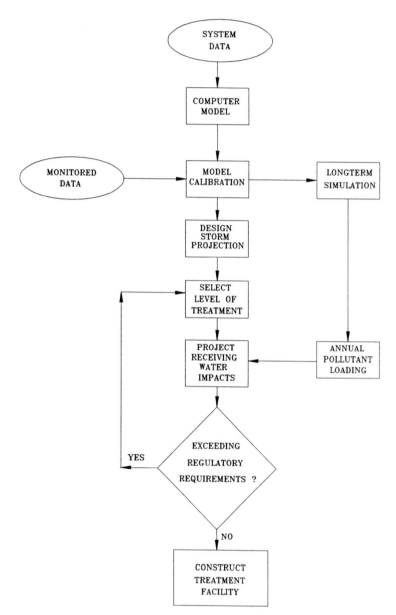

Figure 5.11. Industrial stormwater, project flow chart.

ducted over several months. Sufficient rain gages were utilized to account for rainfall variations over the site. The flow monitoring program utilized a continuous-flow meter at the discharge. Continuous-flow meters were also installed in the receiving stream, upstream and down-stream of the facility. Samples were collected from the discharge and the in-stream monitoring points during several storm events. The samples were collected at short time intervals (e.g., 15 min.) at each of the sampling locations during the first part of each storm event. Additionally, longer time intervals (one hour) were used during the latter part of the events.

Data Summary

Data obtained from the monitoring and sampling programs were used to develop relationships between rainfall intensity, plant discharge, and stream concentrations during the storm events. These relationships were developed both for the discharge and the receiving stream down-stream of the facility. Figures 5.12 and 5.13 illustrate the relationships for the total plant discharge and the receiving stream downstream of the facility, respectively. These relationships show that as flow increased above dry-weather flow, the XYZ concentration in the plant discharge, as well as the receiving water, decreased due to increased flow.

Surface Water Model Development

In an effort to determine how the water quality in the receiving stream reacts to various rainfall scenarios, a surface-water runoff model was developed to predict the flow and associated XYZ concentration. The USEPA's

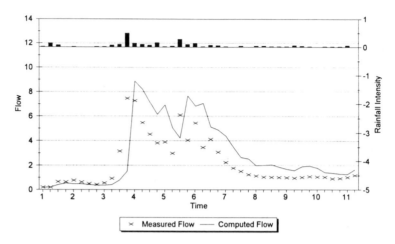

Figure 5.12. Model calibration, total plant discharge.

Storm Water Management Model (SWMM) was used to generate the surface runoff hydrographs from rainfall data.

Data obtained from the flow monitoring program were utilized to calibrate the model. The calibration procedure takes into consideration peak flow rate, flow volume, and

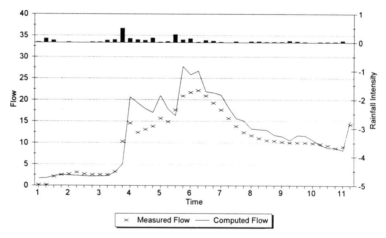

Figure 5.13. Model calibration, receiving stream.

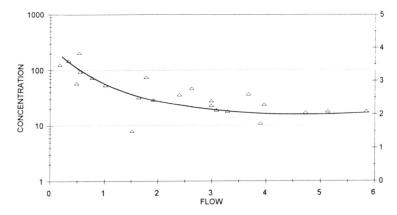

Figure 5.14.   XYZ concentrations vs. flow, total plant discharge.

the shape of the hydrograph for several storm events. The total plant discharge and upstream flows were calibrated against the measured downstream flow. The results of the model calibration for one of these storm events at the plant discharge and receiving stream downstream flow are illustrated by Figures 5.14 and 5.15.

Synthetic rainfall hyetographs were used for the purpose of developing discharge rates for different recurrence-interval storms.

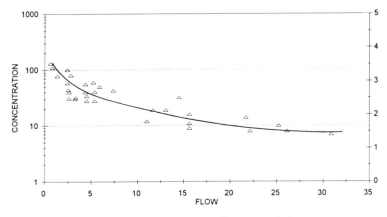

Figure 5.15.   XYZ concentrations vs. flow, receiving stream.

Selecting the Treatment Plant Design Flow

The model was used to project flow and XYZ concentration in the receiving stream immediately downstream of the plant discharge for different rainfall events. The projections for a one-year storm, shown in Figure 5.16, depict the downstream XYZ concentration receding below the permit limit of 11 mg/l during the peak portions of the storm. These modeled results are consistent with the flow and XYZ concentration data collected during two storm events. Flows for a portion of flow condition did not exceed the permit limit, due the effects of dilution. Based on these results, the one-year storm was used as the upper-limit design storm. The quantity of stormwater that required treatment to maintain stream standards was determined by mass-balance calculations. Additionally, consideration of the Best Available Technology (BAT) resulted in stream concentrations well below the standard, as shown on Figure 5.16.

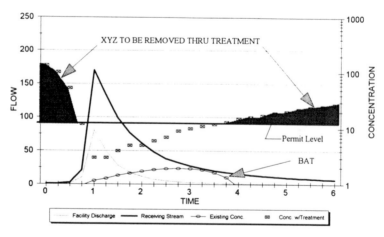

Figure 5.16.   XYZ concentrations vs. flow, receiving water at downstream location, one-year storm.

Continuous simulation was also employed to identify the pollutant reduction that could be expected from the consideration of BAT. Hourly rainfall data were simulated over a typical year and, through mass-balance calculations, it was found that the chosen technology would remove approximately 93% of the annual load of pollutant XYZ.

Summary

Field data and mathematical modeling were used to project the required treatment of an industrial discharge to meet receiving-water quality standards over a range of rainfall conditions. On the basis of this and Best Available Technology (BAT), a treatment plant was designed and constructed, and is operating successfully. The state has accepted the approach as an example methodology to be used for other industrial cases.

6. ONONDAGA LAKE WATERSHED CASE STUDY

Case History By Moffa & Associates

Onondaga Lake, located near Syracuse, New York, has a long history of water-quality problems. It has received discharges from both municipal and industrial point discharges, combined sewer overflows (CSOs), and a diverse mixture of non-point-source (NPS) runoff pollution. Along with CSOs and municipal-treatment plant effluent, NPS pollution is now surfacing as a potentially significant contributor of nutrients, solids, and toxic pollutants to the lake. The extent to which non-point-source pollution can be controlled may determine the extent to which additional CSO abatement may be required.

A modeling approach was undertaken, supplemented

by a spatial analysis utilizing a Geographic Information System (GIS) to estimate the relative impacts of NPS pollution from the urban land areas within the Onondaga Lake Watershed. The spatial analyses was done to quantify the variety of land uses within the watershed. Past studies have shown that the type and quantity of pollutant one might expect from a given area are strongly related to the uses attributed to that area. The hydrology of the watershed was then defined by watershed boundaries, watercourses, slopes, and so on. The pollutant load delivered from each land use was assumed from values developed from other studies. An estimation of the relative impact of runoff-derived pollution was accomplished by combining accurate land use information, hydrology, and pollutant information. This was done by creating a continuous simulation runoff model, using the EPA's Storm Water Management Model (SWMM).

SWMM is a comprehensive urban watershed model with the ability to allow calculation of runoff quantity and quality on a long-term, continuous basis. The EPA's SWMM (RUNOFF block) was used to estimate the annual loads of TSS, total phosphorus, and total nitrogen from the urban portion of each hydrologic unit identified within the watershed. The watershed was divided into 41 distinct hydrologic units (see map, Figure 5.17), each defined by topography as a tributary area. Urban land uses were mapped within each unit and identified, using the GIS software GRASS (Geographical Resources Analysis Support System). GRASS is a GIS developed by the Department of Defense and within the public domain, and is used by a number of federal agencies, including the Soil Conservation Service (SCS).

The SWMM model was used to simulate the hydrologic response of each urban land use within a hydrologic unit. Therefore, some degree of spatial "lumping" was done. The model was simplified by reducing the number of

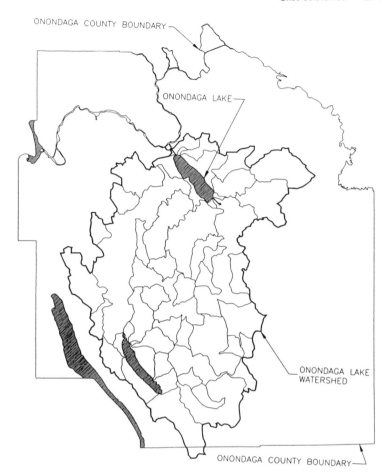

Figure 5.17.   Onondaga Lake watershed, urban land use.

subcatchments. All parcels of a particular land use were lumped together into one catchment area. The hydrologic response was not likely to be affected to an unacceptable degree as a result of this lumping, because the hydrologic units were sufficiently small. Typical input parameters used in the preliminary model are shown in Table 5.2.

The area of each land-use polygon was derived directly from the GIS; the width was calculated from the area, assuming a theoretical rectangle that has a width twice the

Table 5.2. Typical Annual Surface Loads of Pollutants from PLUARG Studies by Marsalek (1978).

Land Use	TSS (lbs/acre)	TP (lbs/acre)	TN (lbs/acre)
Low-density residential	310	1.3	7
High-density residential and commercial	290	2.7	9
Industrial	540	1.8	6
Parks, open space	9	0.03	0.2

breadth. The percent imperviousness was assumed to be directly related to the land use. Therefore, a representative area of each land use was estimated for percent imperviousness, and the resultant appropriate values used for all subcatchments. The slope was derived from the soil information contained in the GIS. Although the slope estimates were somewhat crude, the urban watershed is generally not severely sloped (with some exceptions). The pollutant loads were assumed to be a fraction of the total dust and dirt (milligrams of pollutant over grams of dust and dirt) that builds up over a period of dry days. This conceptual model of the urban watershed was then used to simulate runoff from 30 years of continuous hourly rainfall data. This simulation produced the results shown in Table 5.3.

The values presented are considered reasonable, and compared favorably to literature values for the land uses found in the Onondaga Lake watershed. As of this time, no monitoring has been done to calibrate or validate this simulation. However, the true value of this study is not in the estimation of total annual pollutant load, but in the spatial distribution of the pollutant origin. This has direct impact on the type of possible abatement measures that may be undertaken. For example, if an outstanding

Table 5.3. Estimated Annual Delivered Pollutant Loads and Average Concentrations.

Land Use	TSS (lbs/acre)	TSS (mg/l)	TN (lbs/acre)	TN (mg/l)	TP (lbs/acre)	TP (mg/l)
Low-density residential	290	110	7	2.6	1.2	0.5
High-density residential and commercial	280	70	9	2.2	2.6	0.7
Industrial	520	100	6	1.2	1.7	0.4
Open space	5	8	0.1	0.2	0.02	0.03

amount of pollutants is generated in a concentrated area, then the remedial measures may be different than if the pollutant loads are diffuse in nature and effectively scattered throughout the watershed. Figure 5.18 demonstrates the spatial distribution of TSS intensity (lbs/acre) from urban lands only. This is useful in developing a manage-

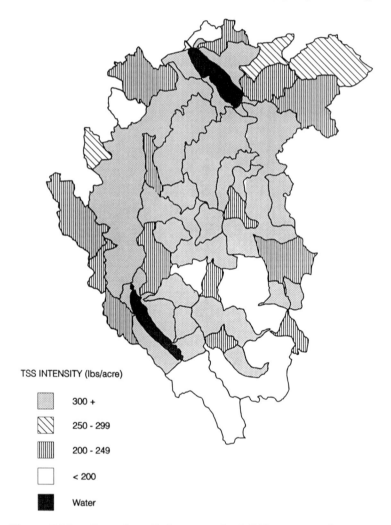

TSS INTENSITY (lbs/acre)

	300 +
	250 - 299
	200 - 249
	< 200
	Water

Figure 5.18.   Onondaga Lake watershed, TSS intensity from urban lands, (lbs/acre).

ment plan for controlling NPS pollution from urban sources.

The results of this study showed that a simplified urban-runoff model predicted reasonable runoff quantities and associated runoff quality. The hydrologic units selected reflected an area large enough to provide for a reasonable-

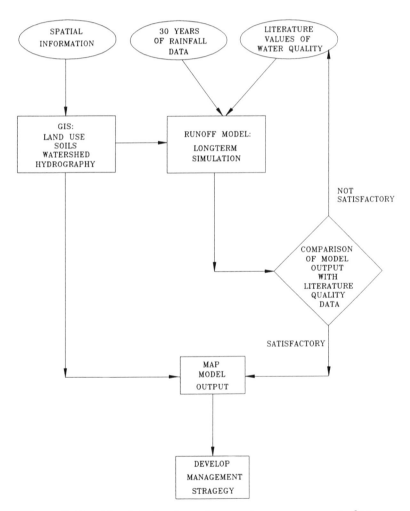

Figure 5.19. Municipal urban stormwater management plan, project flow chart.

sized model, and yet small enough to provide the resultant information at a scale that was useful and suited the goals of the study. The primary goal for this study was to provide a guide in determining a preliminary urban-NPS management plan. A project flow chart for the municipal urban stormwater management plan used in this analysis is provided in Figure 5.19. The relative loads from each hydrologic unit were ranked on the basis of 30 years of continuous simulation. The resultant loads (total pounds of pollutant) were divided by the size of the urban area within each hydrologic unit to provide a guide to the areas of highest pollutant intensity.

The management model can then be used to simulate management scenarios. The cost effectiveness of NPS abatement can then be compared to the costs of CSO abatement of similar pollutants (e.g., phosphorus), thus providing a watershed prospective for abatement and management.

REFERENCES

Townsend, K. L. and Vincent, H. J., "Forging a New Stormwater Strategy," *Industrial Wastewater, Water Environment Federation*, September/October 1994, p. 36.

Appendix:
Outline of
Stormwater-Specific
Federal Register
Documents

1. *Federal Register*, Volume 55, No. 222,
November 16, 1990
National Pollutant Discharge Elimination
System Permit System Application
Regulations for Storm Water Discharges;
Final Rule.

    I. Background of Water Quality Concerns
   II. Water Quality Act of 1987
  III. Remand of 1984 Regulations
  IV. Codification Rule and Case-by-Case Designations
   V. Consent Decree of October 20, 1989
 VI. Today's Final Rule and Response to Comments
      A. Overview
      B. Definition of Storm Water

6. Group Application: Procedural Concerns
7. Permit Applicability and Applications for Oil, Gas and Mining Operations
   a. Gas and Oil Operations
   b. Use of Reportable Quantities to Determine if a Storm Water Discharge from an Oil or Gas Operation is Contaminated
   c. Mining Operations
8. Application Requirements for Construction Activities
   a. Permit Application Requirements
   b. Administrative burdens
G. Municipal Separate Storm Sewer Systems
   1. Municipal Separate Storm Sewers
   2. Effective Prohibition on Non-Storm Water Discharges
   3. Site-Specific Storm Water Quality Management Programs for Municipal Systems
   4. Large and Medium Municipal Storm Sewer Systems
      a. Overview of proposed options and comments
      b. Definition of large and medium
      c. Response to comments
H. Permit Application Requirements for Large and Medium Municipal Systems
   1. Implementing the Permit Program
   2. Structure of Permit Application
      a. Part 1 Application
      b. Part 2 Applications
   3. Major Outfalls
   4. Field Screening Program
   5. Source Identification
   6. Characterization of Discharges
      a. Screening Analysis for Illicit discharges

2. *Federal Register*, Volume 57, No. 64, April 2, 1992.
National Pollutant Discharge Elimination System Application Deadlines, General Permit Requirements and Reporting Requirements for Storm Water Discharges Associated with Industrial Activity; Final Rule.

3. *FEDERAL REGISTER*, VOLUME 57, No. 244, DECEMBER 18, 1992. NATIONAL POLLUTANT DISCHARGE ELIMINATION SYSTEM; STORM WATER DISCHARGES; PERMIT ISSUANCE AND PERMIT COMPLIANCE DEADLINES FOR PHASE I DISCHARGES; FINAL RULE.

4. FEDERAL REGISTER, VOLUME 58, No. 222, NOVEMBER 19, 1993.
WATER POLLUTION CONTROL, NPDES GENERAL PERMITS AND FACT SHEETS: STORM WATER DISCHARGES FROM INDUSTRIAL ACTIVITY; NOTICE.

I. Background
II. Types of Discharges Covered
  A. Limitations on Coverage
III. Pollutants in Storm Water Discharges Associated with Industrial Activities in General
  1. Loading and Unloading Operations
  2. Outdoor Storage
  3. Other Outdoor Activities
  4. Dust or Particulate Generating Processes
  5. Illicit Connections
  6. Waste Management
IV. Summary of Options for Controlling Pollutants
  A. Non-storm Water Discharges
    1. Methods to Identify Non-storm Water Discharges to Separate Storm Sewers
  B. Options for Preventing Pollutants in Storm Water
    1. Elimination of Pollution Sources
    2. Best Management Practices
    3. Traditional Storm Water Management Practices
    4. Diversion of Discharge to Sewage
    5. End-of-Pipe Treatment
V. The Federal/Municipal Partnership: The Role of Municipal Operators of Large and Medium Municipal Separate Storm Sewer Systems
VI. Summary of Common Permit Conditions
  A. Notifications Requirement
    1. Contents of NOIs

P. Motor Freight Transportation Facilities, Passenger Transportation Facilities, Petroleum Bulk Oil Stations and Terminals, Rail Transportation Facilities, and United States Postal Service Transportation Facilities

Q. Water Transportation Facilities That Have Vehicle Maintenance Shops and/or Equipment Cleaning Operations

R. Ship and Boat Building or Repairing Yards

S. Vehicle Maintenance Areas, Equipment Cleaning Areas, or Deicing Areas Located at Air Transportation Facilities

T. Treatment Works

U. Food and Kindred Products Facilities

V. Textile Mills, Apparel, and Other Fabric Product Manufacturing Facilities

W. Wood and Metal Furniture and Fixture Manufacturing Facilities

X. Printing and Publishing Facilities

Y. Rubber, Miscellaneous Plastic Products, and Miscellaneous Manufacturing Industries

Z. Leather Tanning and Finishing Facilities

AA. Fabricated Metal Products Industry

AB. Facilities That Manufacture Transportation Equipment, Industrial, or Commercial Machinery

AC. Facilities That Manufacture Electronic and Electrical Equipment and Components, Photographic and Optical Goods

Glossary

acute toxicity Severe (usually fatal) toxic effects that occur rapidly in affected organisms, due to chemical exposure

advanced treatment Treatment applied for a high degree of pollutant removal in which a significant amount of colloidal and dissolved solids are removed. The process chain generally includes preliminary treatment, secondary biological treatment, and chemical clarification.

advective transport Physical transport of water and associated concentrations from higher to lower hydraulic potential, exclusive of dispersion/mixing

aesthetics Of or pertaining to the sense of attractiveness

algorithm A calculation procedure

aperture size Size of opening between screen material

assimilation In water bodies, the process that removes pollutants and/or their impacts

BAT Best available technology

BCT Best conventional pollutant technology

best management practices (BMP) Measures or practices used to reduce the amount of pollution entering surface waters, air, land, or groundwaters. BMPs may take the form of a process, activity, or physical structure. Some BMPs are simple and can be put into place immediately, while others are more complicated and require extensive planning. BMPs may be inexpensive or costly.

bioaccumulation Accumulation of trace substances in the tissues of organisms

biochemical oxygen demand (BOD) An operational measure of potential for depletion of dissolved oxygen by the biological and chemical degradation of organic material by bacteria

biological treatment processes Means of treatment in which bacterial or biochemical action is intensified to stabilize, oxidize, and nitrify the unstable organic matter present. Trickling filters, activated sludge processes, and lagoons are examples.

biologic diversity A measure of variety of biological organisms

BMP Best management practices

BOD Biochemical oxygen demand

BPJ Best professional judgment

BPT Best practicable control technology

calibration *See* validation

catch basin A chamber or well, usually at the street curbline, for the admission of surface water to a sewer or subdrain, having at its base a sediment sump to retain grit and detritus below the point of overflow; in contrast, a stormwater inlet does not have a sump and does not trap sediment.

catchment The area producing the runoff passing a particular channel or stem location

chlorophyll a A green pigment found in photosynthetic organisms; used as an indicator of algal biomass

chronic toxicity Toxic effects from long-term exposure to chemicals; usually nonfatal to organisms. Effects often consist of reduced growth and reproduction, as well as other physiological impacts

collection system control A method of abating storm-generated or CSO pollution in the collection or drainage system

combined sewer A sewer receiving intercepted surface (dry- and wet-weather) runoff, municipal (sanitary and industrial) sewage, and subsurface waters from infiltration

combined sewer overflow (CSO) Flow from an outfall (discharge conduit) of a combined sewer collection system in excess of the interceptor capacity that is discharged into a receiving-water and/or an auxiliary CSO control (storage) treatment system

composites Water samples comprised of a series of equal-volume samples taken over interval time periods (temporal composite) or spatial locations (spatial composite)

computer model A model in which the mathematical operations are carried out on a computer

continuous simulation The use of a model to simulate the response of a catchment to series of storm events and the hydrologic processes that occur between them

cost–benefit relationship The relationship between unit costs to unit benefits, usually represented as a curve

cost-effective solution A solution to a problem that has been identified as being financially optional (e.g., the solution associated with the knee-of-the-curve of a cost–benefit relationship)

critical design conditions Environmental and flow conditions chosen to represent the conditions under which compliance with water-quality standards, criteria, or objectives is desired

CSO Combined sewer overflow

DAF Dissolved air flotation

desktop analytical models Mathematical models comprised of closed-form solutions that can be readily calculated using a handheld calculator or computer

detention The slowing, dampening, or attenuating of flows either entering the sewer system or within the sewer system by the temporary holding of the water on a surface area, in a storage basin, or within the sewer itself

deterministic mathematical model A mathematical model designed to produce system responses or outputs to temporal and spatial stimuli or inputs

detention time The time period that flow is detained in a storage/sedimentation basin or tank

digital computer model A computer model constructed and operated in the binary representations of digital computers

disinfection The killing or inactivation of human-disease-causing microorganisms or pathogens

dispersion Pollutant or concentration mixing due to turbulent physical processes

dissolved oxygen deficit Difference between saturated dissolved oxygen and ambient concentrations

distributed model A model in which the physical heterogeneities of the catchment are included

diurnal Occurring daily, or over the duration of a day

DMHRF Dual-media high-rate filtration

domestic sewage Sewage derived principally from dwellings, business buildings, institutions, and the like. It may or may not contain groundwater

drogue A floating marker that drifts with the prevailing surface current

dry-weather flow (DWF) Usually referred to as the flow in a combined sewer system without stormwater

dual treatment Those processes or facilities designed for operating on both dry-and-wet-weather flows

DWF Dry-weather flow

dynamic flow routing The movement of water in a channel or over a surface that is modeled with both the continuity-of-mass and momentum equations. This allows the simulation of the propagation of waves (or disturbances) in the upstream and downstream directions.

ecological habitat The environmental niche in which an organism lives

end-of-pipe impacts Impacts that occur in the immediate vicinity of an outfall

enteric Relating to the intestinal tract

eutrophication The process of aging, whereby the increase of mineral and organic nutrients favors aquatic plants over animal life and results in increasing daily variations in dissolved oxygen concentrations, reduced biologic diversity, and reduced water clarity.

event mean concentration (EMC) The arithmetic mean concentration of an urban pollutant measured during a storm runoff event. The EMC is calculated by flow-weighting grab samples or consecutive composite concentrations collected over the course of a storm event.

execution time *See* run time.

FC Fecal coliform

first flush The condition, often occurring in storm-sewer discharges and CSOs, in which a disproportionately high pollutional load is carried to the first portion of the discharge or overflow

floatables Large floating material sometimes characteristic of sanitary wastewater and storm runoff

flow balance method (FBM) A system that stores discharged urban storm runoff in the receiving water for subsequent treatment when capacity is available in the abatement facilities

flow routing The mathematical modeling of flow or flood waves through a channel (e.g., sewer) or over a surface

flux The flow rate per unit cross-sectional area

foul sewer The connecting sewer between the swirl or helical bend regulator/concentrator and the interceptor, that provides both a pathway to the interceptor for sanitary sewage flow during dry-weather flow and concentrated combined sewage during storm-flow periods

frequency of return The rate at which a particular type storm can be expected to occur (e.g., one year), such storms being classified by storm intensity and duration

Geographical Information System (GIS) An information system that combines tabular information with graphical data for efficient collection, storage, retrieval, analysis, and display of spatial data

helical bend regulator/concentrator A CSO control device having an arc or bend shape, providing the same dual functions as cited for the swirl regulator/concentrator

HGMS High-gradient magnetic separation or separator

hydraulic loading The flow rate per unit plan or surface area in sedimentation facilities or cross-sectional area in screening and filtration facilities. *See also* flux.

hydrodynamic   Non-steady-state (dynamic) hydraulic conditions; incremental impacts; fractional portions of an overall impact attributable to different individual sources

industrial stormwater Stormwater discharges associated with industrial activity

infiltration The process whereby water enters a sewer system and service connections underground through such means as, but not limited to, defective pipes, pipe joints, connections, or manhole walls. Infiltration does not include, and is distinguished from, inflow.

initial abstraction The quantity of potential runoff detained by depression storage and initial wetting of the soil

in-line storage A type of storage that has no pumping requirements and can consist of either storage within the sewer (in-sewer/in-pipe) or channel, or storage in the in-line basins

in-pipe storage See in-line storage.

intertidal Related to events during a tidal cycle

intratidal Related to events encompassing a full tidal cycle, or averaged for a tidal cycle

Lagrangian framework A mathematical modeling framework based on a moving reference point in contrast to a fixed coordinate reference point (named after the French mathematician J. L. Lagrange)

lateral A sewer that has no other common sewer discharging into it

longitudinal dispersion One-dimensional dispersion (mixing) occurring along the length of the stream or estuary

lumped model A model in which the physical characteristics of the catchment are assumed to be homogeneous

mathematical modeling Application of mathematical formulae to represent the processes and effects of natural and manmade systems for the purpose of forecasting responses to different conditions and inputs

matrix A fibrous ferromagnetic material in the canister of the high-gradient magnetic separator treatment system

model Any representation of a system by something other than the system itself

model calibration Refinement of mathematical-model parameters and coefficients through comparison to data by application of scientifically consistent and rational adjustments

model parameter A quantity that cannot vary in a particular model run

municipal stormwater Stormwater discharges emanating from residential and commercial areas (e.g., street runoff)

National Pollutant Discharge Elimination System (NPDES) permit. Permit that specifies the requirements of any point source discharge of a pollutant to surface waters

nitrification Biological oxidation of nitrogen compounds by nitrobacteria that consume dissolved oxygen

non-point Diffuse; not attributable to a particular location

non-point-source pollution Any unconfined and non-discrete conveyance from which pollutants are discharged

off-line storage A type of storage that requires detention facilities (basins or tunnels) and facilities for pumping storm flow either into or out of the detention facilities

parameter optimization *See* validation

pathogen A disease-causing microorganism

pathogenic bacteria and viruses Bacteria and viruses capable of causing disease in humans

photosynthesis The process whereby green plants (including algae) convert incident light to chemical energy and synthesize carbohydrates from carbon dioxide and water, with the simultaneous release of oxygen

physical treatment processes Means of treatment in which the application of physical forces predominate. Screening, sedimentation, flotation, and filtration are examples. Physical treatment operations may or may not include chemical additions

point source pollution Pollution that emanates from a well-defined point of discharge which, under most circumstances, is continuous

production run A run of a validated model intended to produce predictive information

RBC Rotating biological contactor

real-time control (RTC) A computerized system that

measures system inputs and monitors stormwater levels in critical sections of the drainage system to retain or release stormwater to maximize storage capacity of the entire drainage system

receiving waters Natural or manmade water systems into which materials are discharged

regulator A structure that controls the amount of sewage entering an interceptor by storing in the upstream trunk line or by diverting some portion of the flow to an outfall

retention The amount of precipitation on a drainage area that does not escape as runoff (the difference between total precipitation and total runoff)

run A single execution of a model

run time The time needed to execute a model run on a computer

Saint-Venant equations The equations of continuity of mass and momentum used to model flow in a channel or over surface

saline intrusion The intruding of saline-rich seawater (which is more dense than fresh water) into a predominantly fresh-water system

sediment oxygen demand Biochemical consumption of dissolved oxygen in overlying waters by decaying sediments across the water–sediment surface

sensitivity analysis The variation of model parameters to determine the sensitivity of the model to each parameter

sewer A pipe or conduit generally closed, and normally not flowing full, for carrying sewage or other waste liquids

sewerage System of piping, with appurtenances, for collecting and conveying wastewaters from source to treatment and/or discharge

s.g. Specific gravity

simulation The representation of physical systems and phenomena by mathematical models

simulation time The time simulated by a model run

single-event simulation The use of a model to simulate the response of a catchment to a single precipitation event (typically a design storm)

source control A method of abating storm-generated or CSO pollution at the upstream, upland source where the pollutants originate and/or accumulate

spatial discretization The level of detail in describing the changes in space of a system, such as a catchment or sewer system, for modeling purposes

SS Suspended solids

storage The slowing, dampening, or attenuating of storm-generated or combined-sewer flows either entering the sewer/drainage system or within the sewer/drainage system, by temporarily holding the flow on a surface area, in a storage basin, within the sewer itself, or within a receiving water

storm flow Overland flow, sewer flow, or receiving-stream flow caused totally or partially by surface storm runoff, storm-related subsurface infiltration, or snowmelt

storm intensity The rate of rainfall, usually expressed as inches per hour

storm sewer A sewer that carries intercepted surface runoff, street wash and other wash waters, or drainage, but excludes domestic sewage and industrial wastes except for unauthorized cross-connections

stormwater Water resulting from precipitation that percolates into the soil, runs off freely from the surface, or is captured by storm sewer/drainage, combined sewer, and, to a limited degree, sanitary sewer facilities

stormwater pollution prevention plan (SPPP) A site-specific plan implemented to minimize and control pollutants in stormwater discharges

stratification Layering of water, caused by differences in density due to temperature and dissolved solids

subcatchment A portion of a catchment producing the runoff that passes a channel or stream location upstream of the location defining the catchment

supernatant The relatively clear liquid layer above the sediment layer in the vertical column

superposition concept Mathematical approach whereby a complex process is represented by the summation of its independent parts

surcharging The transition between open channel flow and pressure flow in sewers

surface runoff Precipitation and water (e.g., street wash) that falls onto the surfaces of roofs, streets, ground, and so on, and is not absorbed or retained by that surface, thereby collecting and running off.

swirl regulator/concentrator A cylindrically shaped CSO control device that provides the dual function of a regulator and a solids–liquid concentrator. As a concentrator, it achieves good removal of the heavier settleable solids fraction in CSO. (*See also* regulator.)

system A network of interacting components capable of responding to one or more stimuli (e.g., watershed, combined sewer system, receiving water)

TC Total coliform

time step The increment of time used to discretely introduce changes in the system

total solids The entire quantity of solids in the liquid flow or volume

toxicity The degree to which a pollutant causes physiological harm to the health of an organism

trace metals Metals present in small concentrations. From a regulatory standpoint, this usually refers to metal concentrations that can cause toxicity at trace concentrations

transient Temporary; of short duration

trunk A sewer, also known as a main sewer, that receives the discharge of one or more submain sewers

urban runoff Surface runoff from an urban drainage area that reaches a stream or other body of water or a sewer/channel

validation The adjustment of model parameters so that the model output adequately mimics the system being modeled

verification The testing of a model to determine whether it is functioning properly

water-quality standard A threshold value or concentration enforced by law as a requirement to maintain acceptable environmental water-quality conditions; usually chosen based on laboratory observations of organism response

watershed management A process whereby all watershed features are evaluated for specific management goals

wetland An area that is periodically inundated or saturated by surface or groundwater on a seasonal or annual basis and that is capable of supporting hydrophytic vegetation. Wetlands typically have hydric soils.

wet-weather flow Usually referred to as the flow in a combined sewer system with stormwater, but may also constitute the flow in a separate storm or sanitary drainage system with stormwater

Index